A FIRST
LOOK
AT
STOCHASTIC
PROCESSES

A FIRST

LOOK

AT STOCHASTIC PROCESSES

Jeffrey S. Rosenthal

University of Toronto, Canada

NEW JERSEY · LONDON · SINGAPORE · BEIJING · SHANGHAI · HONG KONG · TAIPEI · CHENNAI · TOKYO

Published by

World Scientific Publishing Co. Pte. Ltd.

5 Toh Tuck Link, Singapore 596224

USA office: 27 Warren Street, Suite 401-402, Hackensack, NJ 07601

UK office: 57 Shelton Street, Covent Garden, London WC2H 9HE

Library of Congress Cataloging-in-Publication Data

Names: Rosenthal, Jeffrey S. (Jeffrey Seth), author.

Title: A first look at stochastic processes / Jeffrey S. Rosenthal.

Description: New Jersey : World Scientific, [2020] |
    Includes bibliographical references and index.

Identifiers: LCCN 2019044552 | ISBN 9789811207907 (hardcover) |
    ISBN 9789811208973 (paperback)

Subjects: LCSH: Stochastic processes--Textbooks.

Classification: LCC QA274.A529 R67 2020 | DDC 519.2/3--dc23

LC record available at https://lccn.loc.gov/2019044552

British Library Cataloguing-in-Publication Data

A catalogue record for this book is available from the British Library.

For any available supplementary material, please visit
https://www.worldscientific.com/worldscibooks/10.1142/11488#t=suppl

Printed in Singapore

Preface

This book describes the mathematical theory of *stochastic processes*, i.e. of quantities that proceed randomly as a function of time. A main example is Markov chains, which are the focus of the first half of the book and also make frequent appearances in the second half. Simple rules about how processes proceed at each step lead to many surprising, interesting, and elegant theorems about what happens in the long run.

I have tried to communicate the excitement and intrigue of this subject without requiring extensive background knowledge. The book develops a fairly complete mathematical theory of discrete Markov chains and martingales, and then in Section 4 gives some initial ideas about continuous processes. Along the way, it discusses a number of interesting applications, including gambler's ruin, random walks on graphs, sequence waiting times, stock option pricing, branching processes, Markov chain Monte Carlo (MCMC) algorithms, and more.

The book arose from point-form lecture notes. Although it was later expanded into complete sentences and paragraphs, it retains its original brevity, with short paragraphs and each "point" on its own line, to communicate ideas one at a time. Some excessively technical material is *written in smaller font* and may be ignored on first reading. There are also links to some animated simulations; for additional links and updates and information, visit: www.probability.ca/spbook

The target audience for this book is upper-year undergraduate and graduate-level students in Statistics, Mathematics, Computer Science, Economics, Finance, Engineering, Physics, and other subjects which involve logical reasoning and mathematical foundations, and which require working knowledge of how probabilities progress in time. The *prerequisites* to reading this book are a solid background in basic undergraduate-level mathematics, including advanced calculus and basic linear algebra and basic real analysis (*not* including measure theory), plus undergraduate-level probability theory including expected values, distributions, limit theorems, etc.

Appendix A contains many basic facts from elementary probability and mathematics, as needed. It is referred to frequently in the text, to clarify arguments and to fill in any knowledge gaps. Appendix B lists references for further reading about stochastic processes.

Various problems are sprinkled throughout the book, and should be attempted for greater understanding. More involved problems are marked with (*). Problems marked [sol] have full solutions in Appendix C.

To ease understanding and memory, I have provided colorful names for many of the results, like the *Sum Lemma* and *Recurrence Equivalences Theorem* and *Closed Subset Note* and *Vanishing Probabilities Proposition*. Hopefully the names are helpful – otherwise just use their numbers instead. But be warned that if you use these names in conversation, then readers of other books might not know what you are talking about!

Acknowledgements: This text grew out of my lecturing the course STA 447/2006: Stochastic Processes at the University of Toronto over a period of many years. I thank my colleagues for giving me that opportunity. I also thank the many students who have studied these topics with me; their reactions and questions have been a major source of inspiration.

Jeffrey S. Rosenthal
Toronto, Canada, 2019
j.rosenthal@math.toronto.edu
www.probability.ca

About the Author

Jeffrey S. Rosenthal is a Professor of Statistics at the University of Toronto, specialising in Markov chain Monte Carlo (MCMC) algorithms. He received his BSc from the University of Toronto at age 20, and his PhD in Mathematics from Harvard University at age 24. He was awarded the 2006 CRM-SSC Prize, the 2007 COPSS Presidents' Award, the 2013 SSC Gold Medal, and teaching awards at both Harvard and Toronto. He is a fellow of the Institute of Mathematical Statistics and of the Royal Society of Canada. He has published over one hundred research papers and four previous books, including *Struck by Lightning* for the general public which appeared in sixteen editions and ten languages and was a bestseller in Canada, and *A First Look at Rigorous Probability Theory* which presents probability's measure-theoretic mathematical foundations. His website is www.probability.ca, and on Twitter he is @ProbabilityProf.

(Author photo by Henry Chan.)

Contents

1. Markov Chain Probabilities

In this book we will explore the mathematical foundations of *stochastic processes*, i.e. processes that proceed in time in a random way. We will see that they possess many interesting mathematical properties and useful applications. (For background and notation, see Appendix A.)

We begin with one of the most natural and common and interesting types of stochastic processes, namely *Markov chains*.

1.1. First Example: The Frog Walk

Imagine 20 lily pads arranged in a circle.
Suppose a frog starts at pad #20.
Each second, it either stays where it is, or jumps one pad in the clockwise ↻ direction, or jumps one pad in the counter-clockwise ↺ direction, each with probability 1/3. (See Figure 1. Or, for an animated illustration, see www.probability.ca/frogwalk.)

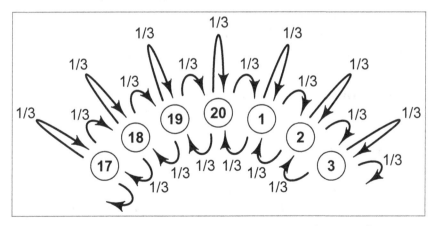

Figure 1: A diagram of part of the Frog Walk example.

This leads to some direct probability computations, e.g.
P(at pad #1 after 1 step) = 1/3.
P(at pad #20 after 1 step) = 1/3.

$\mathbf{P}$(at pad #19 after 1 step) = 1/3.
$\mathbf{P}$(at pad #18 after 1 step) = 0, etc.

Proceeding for <u>two</u> steps, we see that e.g. $\mathbf{P}$(at pad #2 after 2 steps) = $(1/3)(1/3) = 1/9$, since starting at pad #20, there is a 1/3 chance of jumping to pad #1, and then from pad #1 there is a further 1/3 chance of jumping to pad #2.

And, $\mathbf{P}$(at pad #19 after 2 steps) = $(1/3)(1/3) + (1/3)(1/3) = 2/9$, since the frog could *either* first jump from pad #20 to pad #19 and then stay there on the second jump with probability $(1/3)(1/3)$, *or* it could first stay at pad #20 and then jump from there to pad #19 again with probability $(1/3)(1/3)$. And so on.

(1.1.1) Problem. For the Frog Walk, compute $\mathbf{P}$(at pad #1 after 2 steps). **[sol]**   (Note: All problems marked **[sol]** have solutions in Appendix C.)

Markov chain theory concerns itself largely with the question of what happens in the <u>long</u> run.
For example, what is $\mathbf{P}$(frog at pad #14 after 27 steps)?
Farther ahead, what is $\lim_{k\to\infty} \mathbf{P}$(frog at pad #14 after k steps)?
And, will the frog necessarily *eventually* return to pad #20?
Or, will the frog necessarily *eventually* visit every pad?

These questions will all be answered soon in this book.

1.2. Markov Chain Definitions

We begin with a definition.

(1.2.1) Definition. A (discrete-time, discrete-space, time-homogeneous) *Markov chain* is specified by three ingredients:
(i) A *state space* S, any non-empty finite or countable set.
(ii) *initial probabilities* $\{\nu_i\}_{i\in S}$, where ν_i is the probability of starting at i (at time 0). (So, $\nu_i \geq 0$, and $\sum_i \nu_i = 1$.)
(iii) *transition probabilities* $\{p_{ij}\}_{i,j\in S}$, where p_{ij} is the probability of jumping to j if you start at i. (So, $p_{ij} \geq 0$, and $\sum_j p_{ij} = 1$ $\forall i$.)

We will make frequent use of Definition (1.2.1).

In the Frog Walk example:

The state space is $S = \{1, 2, 3, \ldots, 20\}$. (All the lily pads.)
The initial distribution is $\nu_{20} = 1$, and $\nu_i = 0$ for all $i \neq 20$. (Since the frog always starts at pad #20.)
The transition probabilities are given by:

$$p_{ij} = \begin{cases} 1/3, & |j - i| \leq 1 \\ 1/3, & |j - i| = 19 \\ 0, & \text{otherwise} \end{cases}$$

Given <u>any</u> Markov chain, let X_n be the Markov chain's state at time n. So, $X_0, X_1, X_2, \ldots$ are random variables.

Then at time 0, we have $\mathbf{P}(X_0 = i) = \nu_i \ \forall i \in S$.

The transition probabilities p_{ij} can be interpreted as conditional probabilities (A.6.1). Indeed, if $\mathbf{P}(X_n = i) > 0$, then

$$\mathbf{P}(X_{n+1} = j \,|\, X_n = i) = p_{ij}, \quad \forall i, j \in S, \ n = 0, 1, 2, \ldots.$$

(This conditional probability does <u>not</u> depend on n, which is the *time-homogeneous* property.) Also,

$$\mathbf{P}(X_{n+1} = j \,|\, X_0 = i_0, X_1 = i_1, X_2 = i_2, \ldots, X_n = i_n)$$

$$= \mathbf{P}(X_{n+1} = j \,|\, X_n = i_n) = p_{i_n j},$$

i.e. the probabilities at time $n + 1$ depend only on the state at time n (this is called the *Markov property*).

We can then compute joint probabilities by relating them to conditional probabilities. For example, for $i, j \in S$,

$$\mathbf{P}(X_0 = i, X_1 = j) = \mathbf{P}(X_0 = i)\,\mathbf{P}(X_1 = j \,|\, X_0 = i) = \nu_i p_{ij}.$$

Similarly, $\mathbf{P}(X_0 = i, X_1 = j, X_2 = k) = \mathbf{P}(X_0 = i)\,\mathbf{P}(X_1 = j \,|\, X_0 = i)\,\mathbf{P}(X_2 = k \,|\, X_1 = j) = \nu_i p_{ij} p_{jk}$, etc.

More generally,

(1.2.2) $\qquad \mathbf{P}(X_0 = i_0, X_1 = i_1, X_2 = i_2, \ldots, X_n = i_n)$

$$= \nu_{i_0} p_{i_0 i_1} p_{i_1 i_2} \cdots p_{i_{n-1} i_n}.$$

This completely defines the probabilities of the sequence $\{X_n\}_{n=0}^{\infty}$. The random sequence $\{X_n\}_{n=0}^{\infty}$ "is" the Markov chain.

In the Frog Walk example:

At time 0, $\mathbf{P}(X_0 = 20) = \nu_{20} = 1$, and $\mathbf{P}(X_0 = 7) = \nu_7 = 0$, etc.

Also, at time 1, $\mathbf{P}(X_1 = 1) = 1/3$, and $\mathbf{P}(X_1 = 20) = 1/3$, etc.

And, at time 2, as above, $\mathbf{P}(X_2 = 2) = (1/3)(1/3) = 1/9$, and $\mathbf{P}(X_2 = 19) = (1/3)(1/3) + (1/3)(1/3) = 2/9$, etc.

By the formula (1.2.2), we have $\mathbf{P}(X_0 = 20, \ X_1 = 19, \ X_2 = 18) = \nu_{20} p_{20,19} p_{19,18} = (1)(1/3)(1/3) = 1/9$, etc.

(1.2.3) Problem. For the Frog Walk example:

(a) Compute $\mathbf{P}(X_0 = 20, \ X_1 = 19)$. [sol]

(b) Compute $\mathbf{P}(X_0 = 20, \ X_1 = 19, \ X_2 = 20)$. [sol]

(c) Compute $\mathbf{P}(X_0 = 20, \ X_1 = 1, \ X_2 = 2, \ X_3 = 2)$.

(d) Compute $\mathbf{P}(X_0 = 20, \ X_1 = 20, \ X_2 = 19, \ X_3 = 20)$.

(1.2.4) Problem. Repeat Problem (1.2.3) for a *Modified Frog Walk* where each second the frog jumps one pad clockwise ↻ with probability 1/2, or one pad counter-clockwise ↺ with probability 1/3, or stays still with probability 1/6 (see Figure 2).

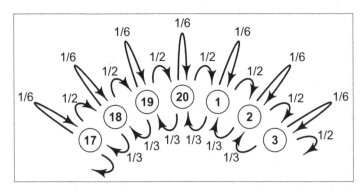

Figure 2: The Modified Frog Walk of Problem (1.2.4).

1.3. Examples of Markov Chains

There are lots of different Markov chains. For example:

(1.3.1) Example. (Simple finite Markov chain)
Let $S = \{1, 2, 3\}$, and $\nu = (1/7, 2/7, 4/7)$, and (see Figure 3)

$$(p_{ij}) = \begin{pmatrix} p_{11} & p_{12} & p_{13} \\ p_{21} & p_{22} & p_{23} \\ p_{31} & p_{32} & p_{33} \end{pmatrix} = \begin{pmatrix} 0 & 1/2 & 1/2 \\ 1/3 & 1/3 & 1/3 \\ 1/4 & 1/4 & 1/2 \end{pmatrix}.$$

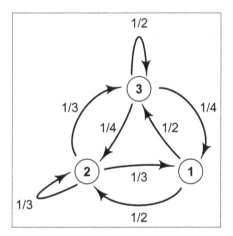

Figure 3: **A diagram of Example (1.3.1).**

Then $\mathbf{P}(X_0 = 3, \ X_1 = 2) = \nu_3 p_{32} = (4/7)(1/4) = 1/7$, and $\mathbf{P}(X_0 = 2, \ X_1 = 1, \ X_2 = 3) = \nu_2 p_{21} p_{13} = (2/7)(1/3)(1/2) = 1/21$, etc.
We can ask, what happens in the long run, as $n \to \infty$?
Is every state visited for sure? What is $\lim_{n \to \infty} \mathbf{P}(X_n = 3)$? etc.

(1.3.2) Problem. For Example (1.3.1):
(a) Compute $\mathbf{P}(X_0 = 1, \ X_1 = 1)$.
(b) Compute $\mathbf{P}(X_0 = 1, \ X_1 = 2)$.
(c) Compute $\mathbf{P}(X_0 = 1, \ X_1 = 1, \ X_2 = 1)$.
(d) Compute $\mathbf{P}(X_0 = 1, \ X_1 = 1, \ X_2 = 1, \ X_3 = 2)$.

(1.3.3) Problem. Consider a Markov chain with $S = \{1, 2, 3\}$, and $\nu = (1/3, 2/3, 0)$, and

$$(p_{ij}) = \begin{pmatrix} 1/2 & 1/3 & 1/6 \\ 1/3 & 0 & 2/3 \\ 0 & 1/4 & 3/4 \end{pmatrix}.$$

(a) Draw a diagram of this Markov chain. [sol]
(b) Compute $\mathbf{P}(X_0 = 1)$.
(c) Compute $\mathbf{P}(X_0 = 1,\ X_1 = 1)$.
(d) Compute $\mathbf{P}(X_0 = 1,\ X_1 = 2)$.
(e) Compute $\mathbf{P}(X_0 = 1,\ X_1 = 1,\ X_2 = 1)$.
(f) Compute $\mathbf{P}(X_0 = 1,\ X_1 = 1,\ X_2 = 1,\ X_3 = 2)$.

(1.3.4) Bernoulli Process. (Like counting sunny days.)
Let $0 < p < 1$. (e.g. $p = 1/2$)
Repeatedly flip a "p-coin" (i.e., a coin with probability of heads $= p$),
at times $1, 2, 3, \ldots$
Let X_n be the number of heads on the first n flips.
Then $\{X_n\}$ is a Markov chain, with $S = \{0, 1, 2, \ldots\}$, $X_0 = 0$ (i.e.
$\nu_0 = 1$, and $\nu_i = 0$ for all $i \neq 0$), and

$$
p_{ij} = \begin{cases} p, & j = i + 1 \\ 1 - p, & j = i \\ 0, & \text{otherwise} \end{cases}
$$

(1.3.5) Problem. For the Bernoulli Process with $p = 1/4$:
(a) Compute $\mathbf{P}(X_0 = 0,\ X_1 = 1)$.
(b) Compute $\mathbf{P}(X_0 = 0,\ X_1 = 1,\ X_2 = 1)$.
(c) Compute $\mathbf{P}(X_0 = 0,\ X_1 = 1,\ X_2 = 1,\ X_3 = 2)$.

(1.3.6) Simple Random Walk (s.r.w.).
Let $0 < p < 1$. (e.g. $p = 1/2$)
Suppose you repeatedly bet \$1.
Each time, you have probability p of winning \$1, and probability
$1 - p$ of losing \$1.
Animated illustration: www.probability.ca/randwalk
Let X_n be the net gain (in dollars) after n bets.
Then $\{X_n\}$ is a Markov chain, with $S = \mathbf{Z}$, and (see Figure 4)

$$
p_{ij} = \begin{cases} p, & j = i + 1 \\ 1 - p, & j = i - 1 \\ 0, & \text{otherwise} \end{cases}
$$

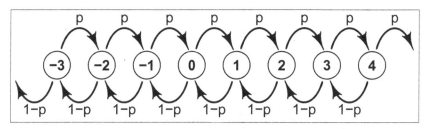

Figure 4: A diagram of Simple Random Walk (s.r.w.).

Usually we take the initial value $X_0 = 0$, so $\nu_0 = 1$. But sometimes we instead take $X_0 = a$ for some other $a \in \mathbf{Z}$, so $\nu_a = 1$.

We can then ask: What happens in the long run? Will you eventually win \$1,000? Will you necessarily eventually go broke? etc.

An important special case is when $p = 1/2$, called *Simple Symmetric Random Walk (s.s.r.w.)* since then $p = 1 - p$.

(1.3.7) Problem. For s.r.w. with $p = 2/3$ and $X_0 = 0$:
(a) Compute $\mathbf{P}(X_0 = 0,\ X_1 = 0)$.
(b) Compute $\mathbf{P}(X_0 = 0,\ X_1 = 1)$.
(c) Compute $\mathbf{P}(X_0 = 0,\ X_1 = 1,\ X_2 = 0)$.
(d) Compute $\mathbf{P}(X_0 = 0,\ X_1 = 1,\ X_2 = 0,\ X_3 = -1)$.

(1.3.8) Ehrenfest's Urn.
We have d balls in total, divided into two urns.
At each time, we choose one of the d balls uniformly at random, and move it to the <u>other</u> urn.
Let X_n be the number of balls in Urn 1 at time n (see Figure 5).
Then $\{X_n\}$ is a Markov chain, with $S = \{0, 1, 2, \ldots, d\}$, and transitions $p_{i,i-1} = i/d$, and $p_{i,i+1} = (d-i)/d$, with $p_{ij} = 0$ otherwise.
(For an animated illustration, see: www.probability.ca/ehrenfest.)
What happens in the long run? Does X_n become uniformly distributed? Does it stay close to X_0? to $d/2$? etc.

(1.3.9) Problem. For Ehrenfest's Urn with $d = 5$ balls and $X_0 = 2$:
(a) Compute $\mathbf{P}(X_0 = 2,\ X_1 = 2)$.
(b) Compute $\mathbf{P}(X_0 = 2,\ X_1 = 1)$.
(c) Compute $\mathbf{P}(X_0 = 2,\ X_1 = 1,\ X_2 = 0)$.
(d) Compute $\mathbf{P}(X_0 = 2,\ X_1 = 1,\ X_2 = 0,\ X_3 = 1)$.

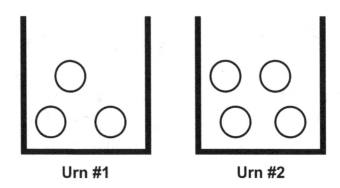

Urn #1   **Urn #2**

Figure 5: Ehrenfest's Urn with $d = 7$ balls, when $X_n = 3$.

(1.3.10) Human Markov Chain.
Each student takes out a coin (or borrows one!).
And, each student selects two other students, one for "Heads" and one for "Tails".
To begin, a stuffed frog is tossed randomly to one student.
Every time the frog comes to a student, that student catches it, flips their coin, and tosses it to their student corresponding to the coin-flip result.
What happens in the long run?
Will every student eventually get the frog?
What about every student who is selected by someone else?
What are the long-run probabilities? etc.

1.4. Multi-Step Transitions

Let $\{X_n\}$ be a Markov chain, with state space S, and transition probabilities p_{ij}, and initial probabilities ν_i. Recall that:
$$\mathbf{P}(X_0 = i_0) = \nu_{i_0}.$$
$$\mathbf{P}(X_0 = i_0,\ X_1 = i_1) = \nu_{i_0} p_{i_0 i_1}.$$
$$\mathbf{P}(X_0 = i_0, X_1 = i_1, \ldots, X_n = i_n) = \nu_{i_0} p_{i_0 i_1} p_{i_1 i_2} \cdots p_{i_{n-1} i_n}.$$

Now, let $\mu_i^{(n)} = \mathbf{P}(X_n = i)$ be the probabilities at time n.
Then at time 0, $\mu_i^{(0)} = \nu_i$.
At time 1, what is $\mu_j^{(1)}$ in terms of ν_i and p_{ij}?

Well, by the *Law of Total Probability* (A.2.3), $\mu_j^{(1)} = \mathbf{P}(X_1 = j) = \sum_{i \in S} \mathbf{P}(X_0 = i, \ X_1 = j) = \sum_{i \in S} \nu_i p_{ij} = \sum_{i \in S} \mu_i^{(0)} p_{ij}$.

These equations can be written nicely with *matrix multiplication* (A.12.1).

Let $m = |S|$ be the number of elements in S (could be infinity).

Write $\nu = (\nu_1, \nu_2, \nu_3, \ldots)$ as a $1 \times m$ row vector.

And, write $\mu^{(n)} = (\mu_1^{(n)}, \mu_2^{(n)}, \mu_3^{(n)}, \ldots)$ as a $1 \times m$ row vector.

And $\mathbf{P} = (p_{ij}) = \begin{pmatrix} p_{11} & p_{12} & p_{13} & \cdots \\ p_{21} & p_{22} & p_{23} & \cdots \\ p_{31} & \vdots & \vdots & \ddots \end{pmatrix}$ as an $m \times m$ matrix.

Then in terms of matrix multiplication, $\mu^{(1)} = \nu P = \mu^{(0)} P$.

Similarly, $\mu_k^{(2)} = \sum_{i \in S} \sum_{j \in S} \nu_i p_{ij} p_{jk}$, etc.

In matrix form: $\mu^{(2)} = \nu PP = \nu P^2 = \mu^{(0)} P^2$.

Then, by induction: $\mu^{(n)} = \nu P^n = \mu^{(0)} P^n$, for $n = 1, 2, 3, \ldots$.

By convention, let $P^0 = I$ be the *identity matrix*.

Then the formula $\mu^{(n)} = \nu P^n$ holds for $n = 0$, too.

For Example (1.3.1), with $S = \{1, 2, 3\}$, and

$$(p_{ij}) = \begin{pmatrix} p_{11} & p_{12} & p_{13} \\ p_{21} & p_{22} & p_{23} \\ p_{31} & p_{32} & p_{33} \end{pmatrix} = \begin{pmatrix} 0 & 1/2 & 1/2 \\ 1/3 & 1/3 & 1/3 \\ 1/4 & 1/4 & 1/2 \end{pmatrix},$$

and $\nu = (1/7, 2/7, 4/7)$, we compute that $\mathbf{P}(X_1 = 2) = \mu_2^{(1)} = \sum_{i \in S} \nu_i p_{i2} = \nu_1 p_{12} + \nu_2 p_{22} + \nu_3 p_{32} = (1/7)(1/2) + (2/7)(1/3) + (4/7)(1/4) = 13/42$, etc.

(1.4.1) Problem. For Example (1.3.1):
(a) Compute $\mathbf{P}(X_1 = 3)$. [sol]
(b) Compute $\mathbf{P}(X_1 = 1)$.
(c) Compute $\mathbf{P}(X_2 = 1)$.
(d) Compute $\mathbf{P}(X_2 = 2)$.
(e) Compute $\mathbf{P}(X_2 = 3)$.

So, for the Frog Walk example, we have e.g. $\mathbf{P}(X_2 = 19) = \mu_{19}^{(2)} = \sum_{i \in S} \sum_{j \in S} \nu_i p_{ij} p_{j,19} = \nu_{20} p_{20,19} p_{19,19} + \nu_{20} p_{20,20} p_{20,19} + 0 = (1)(1/3)(1/3) + (1)(1/3)(1/3) + 0 = 2/9$, just like before.

(1.4.2) Problem. For the Frog Walk example:
(a) Compute $\mathbf{P}(X_1 = 1)$.
(b) Compute $\mathbf{P}(X_1 = 20)$.
(c) Compute $\mathbf{P}(X_2 = 20)$.
(d) Compute $\mathbf{P}(X_2 = 1)$.
(e) Compute $\mathbf{P}(X_2 = 2)$.
(f) Compute $\mathbf{P}(X_2 = 3)$.

(1.4.3) Problem. For the Frog Walk, show $\mathbf{P}(X_3 = 1) = 6/27$.

Another way to track the probabilities of a Markov chain is with *n-step transitions*: $p_{ij}^{(n)} = \mathbf{P}(X_n = j \mid X_0 = i)$.

(Since the chain is *time-homogeneous*, this must be the same as $p_{ij}^{(n)} = \mathbf{P}(X_{m+n} = j \mid X_m = i)$ for any $m \in \mathbf{N}$.)

We must again have $p_{ij}^{(n)} \geq 0$, and $\sum_{j \in S} p_{ij}^{(n)} = \sum_{j \in S} \mathbf{P}_i(X_n = j) = \mathbf{P}_i(X_n \in S) = 1$.

Here $p_{ij}^{(1)} = p_{ij}$. (of course)

But what about, say, $p_{ij}^{(2)}$?

Well, $p_{ij}^{(2)} = \mathbf{P}(X_2 = j \mid X_0 = i) = \sum_{k \in S} \mathbf{P}(X_2 = j, \ X_1 = k \mid X_0 = i) = \sum_{k \in S} p_{ik} p_{kj}$.

In matrix form: $P^{(2)} = \left(p_{ij}^{(2)} \right) = P P = P^2$.

Similarly, $p_{ij}^{(3)} = \sum_{k \in S} \sum_{\ell \in S} p_{ik} p_{k\ell} p_{\ell j}$, i.e. $P^{(3)} = P^3$.

By induction, $P^{(n)} = P^n$ for all $n \in \mathbf{N}$.

That is, to compute probabilities of n-step jumps, you can simply take the nth <u>power</u> of the transition matrix P.

We use the convention that $P^{(0)} = I = $ the identity matrix, i.e.

$$p_{ij}^{(0)} = \delta_{ij} = \begin{cases} 1, & i = j \\ 0, & \text{otherwise} \end{cases}$$

Then $P^{(0)} = P^0$, too, since $P^0 = I$.

(1.4.4) Chapman-Kolmogorov equations. $p_{ij}^{(m+n)} = \sum_{k \in S} p_{ik}^{(m)} p_{kj}^{(n)}$,

and $p_{ij}^{(m+s+n)} = \sum_{k \in S} \sum_{\ell \in S} p_{ik}^{(m)} p_{k\ell}^{(s)} p_{\ell j}^{(n)}$, etc.

Proof: By the *Law of Total Probability* (A.2.3), $p_{ij}^{(m+n)} = \mathbf{P}(X_{m+n} = j \mid X_0 = i) = \sum_{k \in S} \mathbf{P}(X_{m+n} = j, \ X_m = k \mid X_0 = i) = \sum_{k \in S} p_{ik}^{(m)} p_{kj}^{(n)}$.

And, $p_{ij}^{(m+s+n)} = \mathbf{P}(X_{m+s+n} = j \mid X_0 = i) = \sum_{k \in S} \sum_{\ell \in S} \mathbf{P}(X_{m+s+n} = j, X_m = k, X_{m+s} = \ell \mid X_0 = i) = \sum_{k \in S} \sum_{\ell \in S} p_{ik}^{(m)} p_{k\ell}^{(s)} p_{\ell j}^{(n)}.$ ∎

In matrix form, this says: $P^{(m+n)} = P^{(m)} P^{(n)}$, and also $P^{(m+s+n)} = P^{(m)} P^{(s)} P^{(n)}$, etc. (Which also follows from $P^{(n)} = P^n$, since e.g. $P^{(m+n)} = P^{m+n} = P^m P^n = P^{(m)} P^{(n)}$, etc.)

Since all probabilities are ≥ 0, it immediately follows that:

(1.4.5) Chapman-Kolmogorov inequality. $p_{ij}^{(m+n)} \geq p_{ik}^{(m)} p_{kj}^{(n)}$ for any fixed state $k \in S$, and $p_{ij}^{(m+s+n)} \geq p_{ik}^{(m)} p_{k\ell}^{(s)} p_{\ell j}^{(n)}$ for any fixed $k, \ell \in S$, etc.

(1.4.6) Problem. Consider a Markov chain with state space $S = \{1, 2\}$, and transition probabilities $p_{11} = 2/3$, $p_{12} = 1/3$, $p_{21} = 1/4$, and $p_{22} = 3/4$.
(a) Draw a diagram of this Markov chain. [**sol**]
(b) Compute $p_{12}^{(2)}$. [**sol**]
(c) Compute $p_{12}^{(3)}$.

(1.4.7) Problem. For Ehrenfest's Urn with $d = 5$ balls, compute $p_{12}^{(3)}$. [**sol**]

(1.4.8) Problem. Suppose a fair six-sided die is repeatedly rolled, at times $0, 1, 2, 3, \ldots$. (So, each roll is independently equally likely to be 1, 2, 3, 4, 5, or 6.) Let $X_0 = 0$, and for $n \geq 1$ let X_n be the *largest* value that appears among all of the rolls up to time n.
(a) Find (with justification) a state space S, initial probabilities $\{\nu_i\}$, and transition probabilities $\{p_{ij}\}$, for which $\{X_n\}$ is Markov chain.
(b) Compute the two-step transitions $\{p_{35}^{(2)}\}$ and $\{p_{15}^{(2)}\}$.
(c) Compute the three-step transition probability $\{p_{15}^{(3)}\}$.

1.5. Recurrence and Transience

We next consider various return probabilities.

For convenience, we will henceforth sometimes use the *shorthand notation* $\mathbf{P}_i(\cdots)$ for $\mathbf{P}(\cdots \mid X_0 = i)$, and $\mathbf{E}_i(\cdots)$ for $\mathbf{E}(\cdots \mid X_0 = i)$. So, $\mathbf{P}_i(X_n = j) = p_{ij}^{(n)}$, etc.

To continue, let $N(i) = \#\{n \geq 1 : X_n = i\}$ be the total number of times that the chain hits i (not counting time 0).
(So, $N(i)$ is a random variable, possibly infinite, cf. Section A.5.)

And, let f_{ij} be the *return probability* from i to j.
Formally, f_{ij} is defined as $f_{ij} := \mathbf{P}_i(X_n = j$ for some $n \geq 1)$.
That is, f_{ij} is the probability, starting from i, that the chain will eventually visit j at least once.
Equivalently, $f_{ij} = \mathbf{P}_i(N(j) \geq 1)$.
Considering complementary probabilities gives that also:

$$(1.5.1) \qquad 1 - f_{ij} \;=\; \mathbf{P}_i(X_n \neq j \text{ for all } n \geq 1).$$

By stringing different paths together, $\mathbf{P}_i$(chain will eventually visit j, and then eventually visit k) $= f_{ij} f_{jk}$, etc.
Hence, $\mathbf{P}_i(N(i) \geq 2) = (f_{ii})^2$, and $\mathbf{P}_i(N(i) \geq 3) = (f_{ii})^3$, etc.
In general, for $k = 0, 1, 2, \ldots$,

$$(1.5.2) \qquad\qquad \mathbf{P}_i(N(i) \geq k) \;=\; (f_{ii})^k,$$

and similarly

$$(1.5.3) \qquad\qquad \mathbf{P}_i(N(j) \geq k) \;=\; f_{ij}(f_{jj})^{k-1}.$$

It also follows that $f_{ik} \geq f_{ij} f_{jk}$, etc.

(1.5.4) Problem. Let $S = \{1, 2, 3\}$, with $p_{12} = 1/4$, $p_{13} = 3/4$, and $p_{23} = p_{33} = 1$, and $p_{ij} = 0$ otherwise (see Figure 6). Compute f_{12} and f_{13}. [sol]

A very important concept is:

(1.5.5) Definition.   A state i of a Markov chain is *recurrent* (or, *persistent*) if $\mathbf{P}_i(X_n = i$ for some $n \geq 1) = 1$, i.e. if $f_{ii} = 1$.
Otherwise, if $f_{ii} < 1$, then i is *transient*.

We can then ask: Is the Frog Walk recurrent? Is simple random walk? Are the other previous examples?
To help answer these questions, the following result relates recurrence of a state i, to the question of whether the chain will return to i

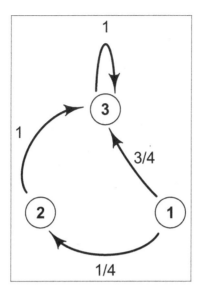

Figure 6: A diagram of the chain in Problem (1.5.4).

infinitely often, and also to whether a certain sum is finite or infinite (for background on infinite series, see Section A.4).

(1.5.6) Recurrent State Theorem.
State i is recurrent $\underline{\text{iff}}$ $\mathbf{P}_i(N(i) = \infty) = 1$ $\underline{\text{iff}}$ $\sum_{n=1}^{\infty} p_{ii}^{(n)} = \infty$.
And, i is transient $\underline{\text{iff}}$ $\mathbf{P}_i(N(i) = \infty) = 0$ $\underline{\text{iff}}$ $\sum_{n=1}^{\infty} p_{ii}^{(n)} < \infty$.

Proof: By continuity of probabilities (A.8.3) and then (1.5.2),

$$\mathbf{P}_i(N(i) = \infty) = \lim_{k \to \infty} \mathbf{P}_i(N(i) \geq k)$$

$$= \lim_{k \to \infty} (f_{ii})^k = \begin{cases} 1, & f_{ii} = 1 \\ 0, & f_{ii} < 1 \end{cases}$$

Then, using countable linearity for non-negative random variables (A.9.2), and then the trick (A.2.1), we have

$$\sum_{n=1}^{\infty} p_{ii}^{(n)} = \sum_{n=1}^{\infty} \mathbf{P}_i(X_n = i) = \sum_{n=1}^{\infty} \mathbf{E}_i(1_{X_n = i})$$

$$= \mathbf{E}_i \left(\sum_{n=1}^{\infty} \mathbf{1}_{X_n=i} \right) = \mathbf{E}_i \left(N(i) \right) = \sum_{k=1}^{\infty} \mathbf{P}_i \left(N(i) \geq k \right)$$

$$= \sum_{k=1}^{\infty} (f_{ii})^k = \begin{cases} \infty, & f_{ii} = 1 \\ \frac{f_{ii}}{1-f_{ii}} < \infty, & f_{ii} < 1 \end{cases} \qquad \blacksquare$$

<u>Aside:</u> The above proof shows that, although f_{ii} is very different from $p_{ii}^{(n)}$, the surprising identity $\sum_{n=1}^{\infty} p_{ii}^{(n)} = \sum_{k=1}^{\infty} (f_{ii})^k$ holds.

What about simple random walk (1.3.6)?
Is the state 0 recurrent in that case?
We'll use the Recurrent State Theorem!
We need to check if $\sum_{n=1}^{\infty} p_{00}^{(n)} = \infty$, or not.
Well, if n is odd, then $p_{00}^{(n)} = 0$.
If n is even, $p_{00}^{(n)} = \mathbf{P}(n/2$ heads and $n/2$ tails on first n tosses).
This is the Binomial(n, p) distribution (A.3.2), so

$$p_{00}^{(n)} = \binom{n}{n/2} p^{n/2}(1-p)^{n/2} = \frac{n!}{[(n/2)!]^2} p^{n/2}(1-p)^{n/2}.$$

But Stirling's Approximation (A.13.3) says that if n is large, then $n! \approx (n/e)^n \sqrt{2\pi n}$. Hence, for n large and even, we also have that $(n/2)! \approx (n/2e)^{n/2}\sqrt{2\pi n/2}$, so therefore

(1.5.7) $$p_{00}^{(n)} \approx \frac{(n/e)^n \sqrt{2\pi n}}{[(n/2e)^{n/2}\sqrt{2\pi n/2}]^2} p^{n/2}(1-p)^{n/2}$$

$$= [4p(1-p)]^{n/2} \sqrt{2/\pi n}.$$

Now, if $p = 1/2$, then $4p(1-p) = 1$, so

$$\sum_{n=1}^{\infty} p_{00}^{(n)} \approx \sum_{n=2,4,6,\ldots} \sqrt{2/\pi n} = \left(\sqrt{2/\pi} \right) \sum_{n=2,4,6,\ldots} n^{-1/2},$$

which is infinite by (A.4.2) with $a = 1/2$.
So, if $p = 1/2$, then state 0 is <u>recurrent</u>, and the chain will return to state 0 infinitely often with probability 1.

But if $p \neq 1/2$, then $4p(1 - p) < 1$, so

$$\sum_{n=1}^{\infty} p_{00}^{(n)} \approx \sum_{n=2,4,6,\ldots} [4p(1 - p)]^{n/2} \sqrt{2/\pi n}$$

$$< \sum_{n=2,4,6,\ldots} [4p(1 - p)]^{n/2} = \frac{4p(1 - p)}{1 - 4p(1 - p)} < \infty$$

since it is a geometric series (A.4.1) with initial term $c = 4p(1 - p)$ and common ratio $r = [4p(1 - p)]^{1/2} < 1$.

So, if $p \neq 1/2$, then state 0 is <u>transient</u>, and the chain will *not* return to state 0 infinitely often.

The same calculation applies to any other state i, proving:

(1.5.8) Proposition. For simple random walk, if $p = 1/2$ then $f_{ii} = 1 \; \forall i$ and all states are recurrent, but if $p \neq 1/2$ then $f_{ii} < 1 \; \forall i$ and all states are transient.

(For a multidimensional version of (1.5.8), see Problem 1.5.15.)

What about f_{ii} for other Markov chains, like the Frog Walk? Coming soon! (See Theorem 1.6.8 below.)

Computing actual values of the return probabilities f_{ij} can be challenging, but there are ways:

(1.5.9) Example. Let $S = \{1, 2, 3, 4\}$, and (see Figure 7)

$$P = \begin{pmatrix} 1 & 0 & 0 & 0 \\ 1/4 & 1/4 & 1/2 & 0 \\ 0 & 0 & 2/5 & 3/5 \\ 0 & 0 & 1/3 & 2/3 \end{pmatrix}.$$

What are the various f_{ij} values?

Here $f_{11} = 1$ (of course).

Also, $f_{22} = 1/4$, since after leaving f it can never return.

Also $f_{33} = 1$, since e.g. $\mathbf{P}_3$(don't return to 3 in first n steps) $= (3/5)(2/3)^{n-1}$ which $\to 0$ as $n \to \infty$. (Alternatively, this can be shown by an infinite sum similar to Solution #2 below.)

By similar reasoning, $f_{44} = 1$.

So, states 1, 3, and 4 are recurrent, but state 2 is transient.

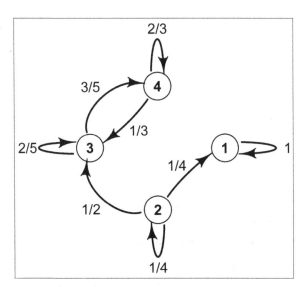

Figure 7: A diagram of Example (1.5.9).

Also, clearly $f_{12} = f_{13} = f_{14} = f_{32} = f_{31} = 0$, since e.g. from state 1 it can never get to state 2, etc.

And, $f_{34} = f_{43} = 1$ (similar to f_{33} above).

What about f_{21}? Is it harder?

To compute the return probabilities f_{ij}, the following result some-times helps:

(1.5.10) f-Expansion. $f_{ij} = p_{ij} + \sum_{\substack{k \in S \\ k \neq j}} p_{ik} f_{kj}$. (Note: The sum does <u>not</u> include the term $k = j$. However, if $i \neq j$, then the sum *does* include the term $k = i$.)

This essentially follows from logical reasoning: from i, to get to j eventually, we have to either jump to j immediately (with probability p_{ij}), or jump to some other state k (with probability p_{ik}) and *then* get to j eventually (with probability f_{kj}).

More formally:

Proof: We have

$$f_{ij} = \mathbf{P}_i(\exists n \geq 1 : X_n = j) = \sum_{k \in S} \mathbf{P}_i(X_1 = k, \exists n \geq 1 : X_n = j)$$

$$= \mathbf{P}_i(X_1 = j, \; \exists n \geq 1 : X_n = j) + \sum_{k \neq j} \mathbf{P}_i(X_1 = k, \; \exists n \geq 1 : X_n = j)$$

$$= p_{ij}(1) + \sum_{k \neq j} p_{ik}(f_{kj}) \, . \qquad \blacksquare$$

For example, the f-Expansion shows $f_{ij} \geq p_{ij}$ (of course).

Returning to Example (1.5.9), what is f_{21}?

Solution #1: By the f-Expansion, $f_{21} = p_{21} + p_{22}f_{21} + p_{23}f_{31} + p_{24}f_{41} = (1/4) + (1/4)f_{21} + (1/2)(0) + (0)(0)$. So, $f_{21} = (1/4) + (1/4)f_{21}$, so $(3/4)f_{21} = (1/4)$, so $f_{21} = (1/4) / (3/4) = 1/3$.

Solution #2: Let τ be the first time we hit 1. Then we compute that $f_{21} = \mathbf{P}_2[\tau < \infty] = \sum_{m=1}^{\infty} \mathbf{P}_2[\tau = m] = \sum_{m=1}^{\infty}(1/4)^{m-1}(1/4) = \sum_{m=1}^{\infty}(1/4)^m = (1/4)/[1 - (1/4)] = 1/3$.

Solution #3: In this special case only, $f_{21} = \mathbf{P}_2(X_1 = 1 \,|\, X_1 \neq 2) = \mathbf{P}_2(X_1 = 1) / \mathbf{P}_2(X_1 \neq 2) = (1/4)/[(1/4) + (1/2)] = 1/3$.

(1.5.11) Problem. In Example (1.5.9):
(a) Show that $f_{23} = 2/3$, in three different ways.
(b) Show that $f_{24} = 2/3$.

(1.5.12) Problem. Consider a Markov chain $\{X_n\}$ on the state space $S = \{1, 2, 3, 4\}$, with transition probabilities

$$(p_{ij}) = \begin{pmatrix} 1/3 & 1/6 & 1/2 & 0 \\ 1/4 & 0 & 3/4 & 0 \\ 2/5 & 3/5 & 0 & 0 \\ 1/2 & 1/2 & 0 & 0 \end{pmatrix}.$$

(a) Compute the value of $p_{43}^{(3)} \equiv \mathbf{P}(X_3 = 3 \,|\, X_0 = 4)$.
(b) Compute f_{23}.
(c) Compute f_{21}.

(1.5.13) Problem. Consider a Markov chain with state space $S = \{1, 2, 3, 4\}$, and transition probabilities:

$$P = \begin{pmatrix} 1/4 & 1/4 & 1/4 & 1/4 \\ 0 & 2/3 & 1/3 & 0 \\ 0 & 1/5 & 4/5 & 0 \\ 0 & 0 & 0 & 1 \end{pmatrix}$$

(a) Draw a diagram of this Markov chain.
(b) Which states are recurrent, and which are transient?
(c) Compute f_{23}.
(d) Compute f_{14}.
(e) Compute f_{13}.

(1.5.14) Problem. For each of the following sets of conditions, either provide an example of Markov chain transition probabilities $\{p_{ij}\}$ on some state space S where the conditions are satisfied, or prove that no such Markov chain exists.
(a) $3/4 < p_{12}^{(n)} < 1$ for all $n \geq 1$.
(b) $p_{11} > 1/2$, and the state 1 is transient.
(c) $p_{12} = 0$ and $p_{12}^{(3)} = 0$, but $0 < p_{12}^{(2)} < 1$.
(d) $f_{12} = 1/2$, and $f_{13} = 2/3$.
(e) $p_{12}^{(n)} \geq 1/4$ and $p_{21}^{(n)} \geq 1/4$ for all $n \geq 1$, and the state 1 is transient.
(f) S is finite, and $\lim\limits_{n \to \infty} p_{ij}^{(n)} = 0$ for all $i, j \in S$.

(g) S is infinite, and $\lim\limits_{n \to \infty} p_{ij}^{(n)} = 0$ for all $i, j \in S$.

(1.5.15) Problem. (*) Consider the Markov chain on $S = \mathbf{Z}^d$ which adds or subtracts 1, with probability $1/2$ each, independently to each of the d coordinates at each step. Thus,

$$P_{(i_1,i_2,\ldots,i_d),\,(j_1,j_2,\ldots,j_d)} = 2^{-d} \quad \text{whenever } |j_r - i_j| = 1 \; \forall \, r.$$

Prove that this chain is recurrent if $d = 1$ or $d = 2$, but transient if $d \geq 3$. [Hint: Use the independence of the coordinates to show that $P_{(0,0,\ldots,0),\,(0,0,\ldots,0)}^{(n)} = \left(\hat{p}_{00}^{(n)}\right)^d$ where $\hat{p}_{ij}$ are the transition probabilities for the usual (one-dimensional) simple symmetric random walk. Then use (1.5.7) together with Theorem (1.5.6) and (A.4.2).]

1.6. Communicating States and Irreducibility

(1.6.1) Definition. State i *communicates* with state j, written $i \to j$, if $f_{ij} > 0$, i.e. if it is possible to get from i to j.

An alternative formulation, which follows easily from countable additivity of probabilities (A.2.2), is:

(1.6.2) $f_{ij} > 0$ iff $\exists m \geq 1$ with $p_{ij}^{(m)} > 0$, i.e. there is <u>some</u> time m for which it is possible to get from i to j in m steps.

We will write $i \leftrightarrow j$ if both $i \to j$ and $j \to i$.

(1.6.3) Definition.   A Markov chain is *irreducible* if $i \to j$ for all $i, j \in S$, i.e. if $f_{ij} > 0 \; \forall i, j \in S$. Otherwise, it is *reducible*.

(1.6.4) Problem. Verify that each of the Frog Walk, and Simple Random Walk, and Ehrenfest's Urn, are all irreducible.

Next, recall the condition $\sum_{n=1}^{\infty} p_{ii}^{(n)} = \infty$ from the Recurrent State Theorem (1.5.6). To check it, the following sometimes helps:

(1.6.5) Sum Lemma. If $i \to k$, and $\ell \to j$, and $\sum_{n=1}^{\infty} p_{k\ell}^{(n)} = \infty$, then $\sum_{n=1}^{\infty} p_{ij}^{(n)} = \infty$.

Proof: By (1.6.2), find $m, r \geq 1$ with $p_{ik}^{(m)} > 0$ and $p_{\ell j}^{(r)} > 0$.
Then by the Chapman-Kolmogorov inequality (1.4.5), $p_{ij}^{(m+s+r)} \geq p_{ik}^{(m)} p_{k\ell}^{(s)} p_{\ell j}^{(r)}$.
Hence, since each $p_{ij}^{(n)} \geq 0$,

$$\sum_{n=1}^{\infty} p_{ij}^{(n)} \geq \sum_{n=m+r+1}^{\infty} p_{ij}^{(n)} = \sum_{s=1}^{\infty} p_{ij}^{(m+s+r)} \geq \sum_{s=1}^{\infty} p_{ik}^{(m)} p_{k\ell}^{(s)} p_{\ell j}^{(r)}$$

$$= p_{ik}^{(m)} p_{\ell j}^{(r)} \sum_{s=1}^{\infty} p_{k\ell}^{(s)} = (positive)(positive)(\infty) = \infty. \quad \blacksquare$$

Setting $j = i$ and $\ell = k$ in the Sum Lemma says that if $i \leftrightarrow k$, then $\sum_{n=1}^{\infty} p_{ii}^{(n)} = \infty$ <u>iff</u> $\sum_{n=1}^{\infty} p_{kk}^{(n)} = \infty$.
Combining this with the Recurrent State Theorem (1.5.6) says:

(1.6.6) Sum Corollary.   If $i \leftrightarrow k$, then i is recurrent <u>iff</u> k is recurrent.

Then, applying this Sum Corollary to an *irreducible* chain (where all states communicate) says:

(1.6.7) Cases Theorem. For an irreducible Markov chain, either
(a) $\sum_{n=1}^{\infty} p_{ij}^{(n)} = \infty$ for all $i, j \in S$, and <u>all</u> states are recurrent. ("*recurrent Markov chain*")
or (b) $\sum_{n=1}^{\infty} p_{ij}^{(n)} < \infty$ for all $i, j \in S$, and <u>all</u> states are transient. ("*transient Markov chain*")

So what about simple random walk (1.3.6)?
Is it irreducible? Yes!
So, the Cases Theorem applies. But which case?
We have already seen (1.5.8) that if $p = 1/2$ then state 0 is recurrent.
Hence, if $p = 1/2$, then it's in case (a), so all states are recurrent, and $\sum_{n=1}^{\infty} p_{ij}^{(n)} = \infty$ for all $i, j \in S$.
Or, if $p \neq 1/2$, then state 0 is transient, so it's in case (b), so all states are transient, and $\sum_{n=1}^{\infty} p_{ij}^{(n)} < \infty$ for all $i, j \in S$.

What about the Frog Walk?
Is it irreducible? Yes!
So, again the Cases Theorem applies. But which case?
The answer is given by:

(1.6.8) Finite Space Theorem. An irreducible Markov chain on a <u>finite</u> state space always falls into case (a), i.e. $\sum_{n=1}^{\infty} p_{ij}^{(n)} = \infty$ for all $i, j \in S$, and all states are recurrent.

Proof: Choose any state $i \in S$.
Then by exchanging the sums (which is valid by (A.9.3) since $p_{ij}^{(n)} \geq 0$), and recalling that $\sum_{j \in S} p_{ij}^{(n)} = 1$, we have

$$\sum_{j \in S} \sum_{n=1}^{\infty} p_{ij}^{(n)} = \sum_{n=1}^{\infty} \sum_{j \in S} p_{ij}^{(n)} = \sum_{n=1}^{\infty} 1 = \infty.$$

This equation holds whether or not S is finite.
But if S is <u>finite</u>, then it follows that there must be at least one $j \in S$ with $\sum_{n=1}^{\infty} p_{ij}^{(n)} = \infty$.

So, we must be in case (a).   ∎

The Frog Walk has $|S| = 20$, which is finite.
So, the Frog Walk is in case (a)!
So, all states are recurrent.
So, in the Frog Walk, $\mathbf{P}_{20}(\exists n \geq 1 \text{ with } X_n = 20) = 1$, etc.
But what about e.g. $f_{20,14} = \mathbf{P}_{20}(\exists n \geq 1 \text{ with } X_n = 14)$?

To solve such problems, for $i \neq j$ let H_{ij} be the event that the chain hits the state i before returning to j, i.e.

$$H_{ij} = \{\exists n \in \mathbf{N} : X_n = i, \text{ but } X_m \neq j \text{ for } 1 \leq m \leq n - 1\}.$$

(1.6.9) Hit Lemma. If $j \to i$ with $j \neq i$, then $\mathbf{P}_j(H_{ij}) > 0$. (That is: If it is possible to get from j to i at all, then it is possible to get from j to i without first returning to j.)

The Hit Lemma is intuitively obvious: If there is some path from j to i, then the final part of the path (starting with the last time it visits i) is a possible path from j to i which does not return to i.

For a more formal proof:
Since $j \to i$, there is some possible path from j to i, i.e. there is $m \in \mathbf{N}$ and $x_0, x_1, \ldots, x_m$ with $x_0 = j$ and $x_m = i$ and $p_{x_r x_{r+1}} > 0$ for all $0 \leq r \leq m - 1$.
Let $S = \max\{r : x_r = j\}$ be the last time this path hits j. (So, we might have $S = 0$, but might have $S > 0$.)
Then $x_S, x_{S+1}, \ldots, x_m$ is a possible path which goes from j to i without first returning to j.
So, $\mathbf{P}_j(H_{ij}) \geq \mathbf{P}_j(\text{this path}) = p_{x_S x_{S+1}} p_{x_{S+1} x_{S+2}} \cdots p_{x_{m-1} x_m} > 0$, thus giving the result.   ∎

Using the Hit Lemma, we can prove:

(1.6.10) f-Lemma. If $j \to i$ and $f_{jj} = 1$, then $f_{ij} = 1$.

Proof: If $i = j$ it's trivial, so assume $i \neq j$.
Since $j \to i$, we have $\mathbf{P}_j(H_{ij}) > 0$ by the Hit Lemma.
But one way to never return to j, is to first hit i, and then from i never return to j.
That is, $\mathbf{P}_j(\text{never return to } j) \geq \mathbf{P}_j(H_{ij}) \mathbf{P}_i(\text{never return to } j)$.

By (1.5.1), this means $1 - f_{jj} \geq \mathbf{P}_j(H_{ij})(1 - f_{ij})$.
But if $f_{jj} = 1$, then $1 - f_{jj} = 0$, so $\mathbf{P}_j(H_{ij})(1 - f_{ij}) = 0$.
Since $\mathbf{P}_j(H_{ij}) > 0$, we must have $1 - f_{ij} = 0$, i.e. $f_{ij} = 1$.　■

This also allows us to prove:

(1.6.11) Infinite Returns Lemma. For an irreducible Markov chain, if it is recurrent then $\mathbf{P}_i(N(j) = \infty) = 1 \ \forall i, j \in S$, but if it is transient then $\mathbf{P}_i(N(j) = \infty) = 0 \ \forall i, j \in S$.

Proof: If the chain is recurrent, then $f_{ij} = f_{jj} = 1$ by the f-Lemma. Hence, similar to the proof of the Recurrent State Theorem (1.5.6), again using continuity of probabilities (A.8.3) and then the formula (1.5.3) for $\mathbf{P}_i(N(j) \geq k)$, we have

$$\mathbf{P}_i(N(j) = \infty) = \lim_{k \to \infty} \mathbf{P}_i(N(j) \geq k)$$

$$= \lim_{k \to \infty} f_{ij}(f_{jj})^{k-1} = \lim_{k \to \infty} (1)(1)^{k-1} = 1.$$

Or, if instead the chain is transient, then $f_{jj} < 1$, so $\mathbf{P}_i(N(j) = \infty) = \lim_{k \to \infty} f_{ij}(f_{jj})^{k-1} = 0$.　■

Putting all of the above together, we obtain:

(1.6.12) Recurrence Equivalences Theorem. If a chain is irreducible, the following are equivalent (all in "case (a)"):
(1)　There are $k, \ell \in S$ with $\sum_{n=1}^{\infty} p_{k\ell}^{(n)} = \infty$.
(2)　For all $i, j \in S$, we have $\sum_{n=1}^{\infty} p_{ij}^{(n)} = \infty$.
(3)　There is $k \in S$ with $f_{kk} = 1$, i.e. with k recurrent.
(4)　For all $j \in S$, we have $f_{jj} = 1$, i.e. all states are recurrent.
(5)　For all $i, j \in S$, we have $f_{ij} = 1$.
(6)　There are $k, \ell \in S$ with $\mathbf{P}_k(N(\ell) = \infty) = 1$.
(7)　For all $i, j \in S$, we have $\mathbf{P}_i(N(j) = \infty) = 1$.

Proof: All of the necessary implications follow from results that we have already proven:
$(1) \Rightarrow (2)$: Sum Lemma.
$(2) \Rightarrow (4)$: Recurrent State Theorem (with $i = j$).

$(4) \Rightarrow (5)$: f-Lemma.

$(5) \Rightarrow (3)$: Immediate.

$(3) \Rightarrow (1)$: Recurrent State Theorem (with $\ell = k$).

$(4) \Rightarrow (7)$: Infinite Returns Lemma.

$(7) \Rightarrow (6)$: Immediate.

$(6) \Rightarrow (3)$: Infinite Returns Lemma.   ∎

Or, considering the <u>opposites</u> (and (1.6.11)), we obtain:

(1.6.13) Transience Equivalences Theorem. If a chain is irreducible, the following are equivalent (all in "case (b)"):

(1)   For all $k, \ell \in S$, $\sum_{n=1}^{\infty} p_{k\ell}^{(n)} < \infty$.

(2)   There is $i, j \in S$ with $\sum_{n=1}^{\infty} p_{ij}^{(n)} < \infty$.

(3)   For all $k \in S$, $f_{kk} < 1$, i.e. k is transient.

(4)   There is $j \in S$ with $f_{jj} < 1$, i.e. some state is transient.

(5)   There are $i, j \in S$ with $f_{ij} < 1$.

(6)   For all $k, \ell \in S$, $\mathbf{P}_k(N(\ell) = \infty) = 0$.

(7)   There are $i, j \in S$ with $\mathbf{P}_i(N(j) = \infty) = 0$.

What about the Frog Walk?

The Finite Space Theorem (1.6.8) says it's in case (a).

Then, the Recurrence Equivalences Theorem (1.6.12) says $f_{ij} = 1$ for all $i, j \in S$.

So, $\mathbf{P}(\exists n \geq 1 \text{ with } X_n = 14 \,|\, X_0 = 20) = f_{20,14} = 1$, etc.

What about simple symmetric ($p = 1/2$) random walk?

We know from (1.5.8) that it's in case (a).

So, the Recurrence Equivalences Theorem (1.6.12) says $f_{ij} = 1$ for all $i, j \in S$.

Hence, $\mathbf{P}(\exists n \geq 1 \text{ with } X_n = 1,000,000 \,|\, X_0 = 0) = 1$, etc.

More generally, for any sequence $i_1, i_2, i_3, \ldots \in S$,

$$(1.6.14) \qquad\qquad f_{i_1 i_2} f_{i_2 i_3} f_{i_3 i_4} \cdots \; = \; 1\,.$$

This means that for any conceivable pattern of values, with probability 1 the chain will eventually hit each of them, in sequence.

That is, the chain has infinite *fluctuations*.

(1.6.15) Example. Let $S = \{1, 2, 3\}$, and

$$(p_{ij}) = \begin{pmatrix} 0 & 1/2 & 1/2 \\ 0 & 1 & 0 \\ 0 & 0 & 1 \end{pmatrix}.$$

Then of course $f_{22} = 1$ (since $p_{22} = 1$), so state 2 is recurrent.
Also $f_{11} = 0 < 1$, so state 1 is transient.
Also, for all $n \in \mathbf{N}$, $p_{12}^{(n)} = 1/2$.
So, $\sum_{n=1}^{\infty} p_{12}^{(n)} = \sum_{n=1}^{\infty} (1/2) = \infty$.
Also, for all $n \in \mathbf{N}$, $p_{21}^{(n)} = 0$.
So, $\sum_{n=1}^{\infty} p_{12}^{(n)} = \sum_{n=1}^{\infty} 0 = 0 < \infty$.
Also, $f_{12} = 1/2 < 1$.
In particular, some states are recurrent, and some transient.
And, sometimes $f_{ij} = 1$, but sometimes $f_{ij} < 1$.
And, sometimes $\sum_n p_{ij}^{(n)} = \infty$, but sometimes $\sum_n p_{ij}^{(n)} < \infty$.
Are these contradictions to the Recurrence Equivalence Theorem?
No, since the chain is reducible, i.e. not irreducible.

One observation which sometimes helps is:

(1.6.16) Closed Subset Note. Suppose a chain is reducible, but
has a *closed subset* $C \subseteq S$ (i.e., $p_{ij} = 0$ for $i \in C$ and $j \notin C$) on which
it is irreducible (i.e., $i \rightarrow j$ for all $i, j \in C$). Then, the Recurrence
Equivalences Theorem, and all results about irreducible chains, still
apply to the chain *restricted* to C.

To illustrate this, consider again Example (1.5.9) with

$$P = \begin{pmatrix} 1 & 0 & 0 & 0 \\ 1/4 & 1/4 & 1/2 & 0 \\ 0 & 0 & 2/5 & 3/5 \\ 0 & 0 & 1/3 & 2/3 \end{pmatrix}.$$

Is this chain irreducible? No!
But it is irreducible on the closed subset $C = \{3, 4\}$.
Since C is finite, the Finite Space Theorem (1.6.8) says that the chain
restricted to C is in case (a).

This immediately proves that states 3 and 4 are recurrent, and that $f_{34} = f_{43} = 1$, and $\sum_{n=1}^{\infty} p_{43}^{(n)} = \infty$, etc.
Easier than before!

Now, consider simple random walk with $p > 1/2$ (e.g. $p = 0.51$). Is it irreducible? Yes!
Is $f_{00} = 1$? No, $f_{00} < 1$ (transient).   Similarly $f_{55} < 1$, etc.
By the Cases Theorem (1.6.7), $\sum_{n=1}^{\infty} p_{ij}^{(n)} < \infty$ for all $i, j \in S$.
On the other hand, surprisingly, we still have e.g. $f_{05} = 1$:

(1.6.17) Proposition. For simple random walk with $p > 1/2$, $f_{ij} = 1$ whenever $j > i$. (Or, if $p < 1/2$ and $j < i$, then $f_{ij} = 1$.)

Proof: Let $X_0 = 0$, and $Z_n = X_n - X_{n-1}$ for $n = 1, 2, \ldots$.
Then, by construction, $X_n = \sum_{i=1}^{n} Z_i$.
The $\{Z_n\}$ are i.i.d., with $\mathbf{P}(Z_n = +1) = p$ and $\mathbf{P}(Z_n = -1) = 1 - p$.
So, by the Law of Large Numbers (A.7.2), w.p. 1, $\lim_{n\to\infty} \frac{1}{n}(Z_1 + Z_2 + \ldots + Z_n) = \mathbf{E}(Z_1) = p(1) + (1-p)(-1) = 2p - 1 > 0$.
So, w.p. 1, $\lim_{n\to\infty}(Z_1 + Z_2 + \ldots + Z_n) = +\infty$.
So, w.p. 1, $X_n - X_0 \to \infty$, i.e. $X_n \to \infty$.
So, starting from i, the chain will converge to ∞.
If $i < j$, then to go from i to ∞, the chain must pass through j.
This will happen with probability 1, so $f_{ij} = 1$. ∎

For example, if $p = 0.51$ and $i = 0$ and $j = 5$, then $f_{05} = 1$.
Does this contradict the Recurrence Equivalences Theorem?
No! We could still have $\sum_{n=1}^{\infty} p_{05}^{(n)} < \infty$, etc.
This is why the Recurrence Equivalences Theorem does <u>not</u> include the equivalence "There are $k, \ell \in S$ with $f_{k\ell} = 1$".

(1.6.18) Problem. Consider the Markov chain from Problem (1.4.6), with state space $S = \{1, 2\}$, and transition probabilities $p_{11} = 2/3$, $p_{12} = 1/3$, $p_{21} = 1/4$, and $p_{22} = 3/4$.
(a) Determine whether or not this chain is irreducible. [sol]
(b) Determine whether or not $f_{11} = 1$. [sol]
(c) Determine whether or not $f_{21} = 1$. [sol]
(d) Determine whether or not $\sum_{n=1}^{\infty} p_{11}^{(n)} = \infty$. [sol]
(e) Determine whether or not $\sum_{n=1}^{\infty} p_{21}^{(n)} = \infty$. [sol]

(1.6.19) Problem. Let $S = \{1, 2, 3\}$, with transition probabilities defined by $p_{11} = p_{12} = p_{22} = p_{23} = p_{32} = p_{33} = 1/2$, with $p_{ij} = 0$ otherwise.
(a) Is this chain irreducible?   [**sol**]
(b) Compute f_{ii} for each $i \in S$.   [**sol**]
(c) Specify which states are recurrent, and which states are transient. [**sol**]
(d) Compute the value of f_{13}.   [**sol**]

(1.6.20) Problem.   For each of the following sets of conditions, either provide (with explanation) an example of a state space S and Markov chain transition probabilities $\{p_{ij}\}_{i,j \in S}$ such that the conditions are satisfied, or prove that no such Markov chain exists.
(a) The chain is irreducible and transient, and there are $k, \ell \in S$ with $p_{k\ell}^{(n)} \geq 1/3$ for all $n \in \mathbf{N}$.   [**sol**]
(b) The chain is irreducible, and there are distinct states $i, j, k, \ell \in S$ such that $f_{ij} < 1$, and $\sum_{n=1}^{\infty} p_{k\ell}^{(n)} = \infty$.   [**sol**]
(c) There are distinct states $k, \ell \in S$ such that if the chain starts at k, then there is a <u>positive</u> probability that the chain will visit ℓ exactly <u>five</u> times (and then never again).   [**sol**]
(d) The chain is irreducible and transient, and there are $k, \ell \in S$ with $f_{k\ell} = 1$.   [**sol**]

(1.6.21) Problem.   Consider the chain from Problem (1.5.12), with $S = \{1, 2, 3, 4\}$ and

$$(p_{ij}) = \begin{pmatrix} 1/3 & 1/6 & 1/2 & 0 \\ 1/4 & 0 & 3/4 & 0 \\ 2/5 & 3/5 & 0 & 0 \\ 1/2 & 1/2 & 0 & 0 \end{pmatrix}.$$

Use (1.6.16) to compute f_{21} and f_{23} more easily.

(1.6.22) Problem. Consider a Markov chain with state space $S = \{1, 2, 3, 4\}$, and transition probabilities $p_{11} = p_{12} = 1/2$, $p_{21} = 1/3$, $p_{22} = 2/3$, $p_{32} = 1/7$, $p_{33} = 2/7$, $p_{34} = 4/7$, $p_{44} = 1$, with $p_{ij} = 0$ otherwise.
(a) Draw a diagram of this Markov chain.

(b) Compute $p_{32}^{(2)}$. [sol]

(c) Determine whether or not $\sum_{n=1}^{\infty} p_{12}^{(n)} = \infty$. [Hint: Perhaps let $C = \{1, 2\}$.] [sol]

(d) Compute f_{32}, in three different ways. [sol]

(1.6.23) Problem. Consider a Markov chain with state space $S = \{1, 2, 3\}$, and transition probabilities $p_{11} = 1/6$, $p_{12} = 1/3$, $p_{13} = 1/2$, $p_{22} = p_{33} = 1$, and $p_{ij} = 0$ otherwise.

(a) Draw a diagram of this Markov chain.

(b) Compute (with explanation) f_{12}.

(c) Prove that $p_{12}^{(n)} \geq 1/3$, for all positive integers n.

(d) Compute $\sum_{n=1}^{\infty} p_{12}^{(n)}$.

(e) Why do the answers in parts (d) and (b) not contradict the implication (1) $\Rightarrow$ (5) in the Recurrence Equivalences Theorem?

(1.6.24) Problem. Consider a Markov chain with $S = \{1, 2, 3, 4, 5, 6, 7\}$, and transition probabilities

$$
(p_{ij}) = \begin{pmatrix}
1 & 0 & 0 & 0 & 0 & 0 & 0 \\
1/2 & 0 & 1/2 & 0 & 0 & 0 & 0 \\
0 & 1/5 & 4/5 & 0 & 0 & 0 & 0 \\
0 & 0 & 1/5 & 2/5 & 2/5 & 0 & 0 \\
1/10 & 0 & 0 & 0 & 7/10 & 0 & 1/5 \\
0 & 0 & 0 & 0 & 0 & 0 & 1 \\
0 & 0 & 0 & 0 & 0 & 1 & 0
\end{pmatrix}
$$

(a) Draw a diagram of this Markov chain.

(b) Determine which states are recurrent, and which are transient.

(c) Compute f_{i1} for each $i \in S$. (Hint: Leave f_{41} until last.)

(1.6.25) Problem. For each of the following sets of conditions, either provide (with explanation) an example of a state space S and Markov chain transition probabilities $\{p_{ij}\}_{i,j \in S}$ such that the conditions are satisfied, or prove that no such Markov chain exists.

(a) There are $i, j \in S$ with $0 < f_{ij} < 1$, and $p_{ij}^{(n)} = 0$ for all $n \geq 3$.

(b) There are distinct states $i, j \in S$ with $f_{ij} > 0$ and $f_{ji} > 0$, and i

is transient.

(c) The chain is irreducible, and there are distinct states $i, j, k, \ell \in S$ such that $f_{ij} = 1$, and $\sum_{n=1}^{\infty} p_{k\ell}^{(n)} < \infty$.

(d) There are distinct states $i, j, k \in S$ with $i \leftrightarrow j$ and $k \to i$, and $\sum_{n=1}^{\infty} p_{kj}^{(n)} < \infty$.

(e) The state space $S = \{1, 2, 3\}$, and $1 \leftrightarrow 2$, and $3 \to 1$, and $\sum_{n=1}^{\infty} p_{32}^{(n)} < \infty$.

1.7. Application – Gambler's Ruin

Consider the following gambling game.

Let $0 < a < c$ be integers, and let $0 < p < 1$.

Suppose player A starts with a dollars, and player B starts with $c - a$ dollars, and they repeatedly bet.

At each bet, A wins \$1 from B with probability p, or B wins \$1 from A with probability $1 - p$.

If X_n is the amount of money that A has at time n, then clearly $X_0 = a$, and $\{X_n\}$ follows a simple random walk (1.3.6).

Let $T_i = \inf\{n \geq 0 : X_n = i\}$ be the first time A has i dollars.

The *Gambler's Ruin* question is: what is $\mathbf{P}_a(T_c < T_0)$, i.e. what is the probability that A reaches c dollars <u>before</u> reaching 0 (i.e., before losing all their money)?

For an animated version, see: www.probability.ca/gambler

Some specific examples are:

What is the probability if $c = 10,000$, $a = 9,700$, and $p = 0.5$?

What if instead $c = 10,000$, $a = 9,700$, and $p = 0.49$?

Is the probability higher if $c = 8$, $a = 6$, $p = 1/3$ ("born rich"), or if $c = 8$, $a = 2$, $p = 2/3$ ("born lucky")?

Here $\{X_n\}$ is a Markov chain.

However, there is no limit to how long the game might take.

So, how can we find this win probability $\mathbf{P}_a(T_c < T_0)$?

Key: Write $\mathbf{P}_a(T_c < T_0)$ as $s(a)$, and consider it to be a <u>function</u> of the player's initial fortune a.

Can we relate the different unknown $s(a)$ to each other? Yes!

Clearly $s(0) = 0$, and $s(c) = 1$.

Furthermore, on the <u>first</u> bet, A either wins or loses $1.
So, for $1 \leq a \leq c - 1$, by conditioning (A.6.5),

$$s(a) = \mathbf{P}_a(T_c < T_0)$$

$$= \mathbf{P}_a(T_c < T_0, \ X_1 = X_0 + 1) + \mathbf{P}_a(T_c < T_0, \ X_1 = X_0 - 1)$$

$$= \mathbf{P}(X_1 = X_0 + 1) \, \mathbf{P}_a(T_c < T_0 \,|\, X_1 = X_0 + 1)$$

$$+ \mathbf{P}(X_1 = X_0 - 1) \, \mathbf{P}_a(T_c < T_0 \,|\, X_1 = X_0 - 1)$$

$$= p \, s(a + 1) + (1 - p) \, s(a - 1).$$

This gives $c - 1$ equations for the $c - 1$ unknowns.

We can solve this using simple algebra!
(It can also be solved using the theory of *difference equations*.)
Re-arranging, $p \, s(a) + (1 - p) \, s(a) = p \, s(a + 1) + (1 - p) \, s(a - 1)$.
Hence, $s(a + 1) - s(a) = \frac{1-p}{p} [s(a) - s(a - 1)]$.
Let $x = s(1)$ (an unknown quantity).
Then $s(1) - s(0) = x$, and $s(2) - s(1) = \frac{1-p}{p}[s(1) - s(0)] = \frac{1-p}{p} x$.
Then $s(3) - s(2) = \frac{1-p}{p}[s(2) - s(1)] = \left(\frac{1-p}{p}\right)^2 x$.
In general, for $1 \leq a \leq c - 1$, $s(a + 1) - s(a) = \left(\frac{1-p}{p}\right)^a x$.
Hence, for $1 \leq a \leq c - 1$,

$$s(a) = s(a) - s(0)$$

$$= [s(a) - s(a - 1)] + [s(a - 1) - s(a - 2)] + \ldots + [s(1) - s(0)]$$

$$= \left[\left(\frac{1 - p}{p}\right)^{a-1} + \left(\frac{1 - p}{p}\right)^{a-2} + \ldots + \left(\frac{1 - p}{p}\right) + 1 \right] x$$

$$= \begin{cases} \left[\frac{\left(\frac{1-p}{p}\right)^a - 1}{\left(\frac{1-p}{p}\right) - 1} \right] x, & p \neq 1/2 \\ \\ ax, & p = 1/2 \end{cases}$$

But $s(c) = 1$, so we can solve for x:

$$x = \begin{cases} \dfrac{\left(\frac{1-p}{p}\right) - 1}{\left(\frac{1-p}{p}\right)^c - 1}, & p \neq 1/2 \\ \\ 1/c, & p = 1/2 \end{cases}$$

We then obtain our final *Gambler's Ruin formula*:

$$(1.7.1) \qquad s(a) = \begin{cases} \dfrac{\left(\frac{1-p}{p}\right)^a - 1}{\left(\frac{1-p}{p}\right)^c - 1}, & p \neq 1/2 \\ \\ a/c, & p = 1/2 \end{cases}$$

thus finally solving the Gambler's Ruin problem as desired.

Some specific examples are:
If $c = 10,000$, $a = 9,700$, and $p = 0.5$, then

$$s(a) = a/c = 0.97.$$

Or, if $c = 10,000$ and $a = 9,700$, but $p = 0.49$, then

$$s(a) = \frac{\left(\frac{0.51}{0.49}\right)^{9,700} - 1}{\left(\frac{0.51}{0.49}\right)^{10,000} - 1} \doteq 0.000006134 \doteq \frac{1}{163,000}.$$

So, changing p from 0.5 to 0.49 makes a huge difference!
Also, if $c = 8$, $a = 6$, and $p = 1/3$ ("born rich"), then

$$s(a) = \frac{\left(\frac{2/3}{1/3}\right)^6 - 1}{\left(\frac{2/3}{1/3}\right)^8 - 1} = \frac{63}{255} \doteq 0.247,$$

but if $c = 8$, $a = 2$, and $p = 2/3$ ("born lucky"), then

$$s(a) = \frac{\left(\frac{1/3}{2/3}\right)^2 - 1}{\left(\frac{1/3}{2/3}\right)^8 - 1} = \frac{3/4}{255/256} \doteq 0.753.$$

So, in this case, it is better to be born lucky than rich!

We will sometimes write $s(a)$ as $s_{c,p}(a)$, to show the explicit dependence on c and p.

(1.7.2) Problem. (*) [Calculus Challenge] Is $s_{c,p}(a)$ a continuous function of p, as $p \to 1/2$? That is, does $\lim_{p \to 1/2} s_{c,p}(a) = s_{c,1/2}(a)$? [Hint: Don't forget *L'Hôpital's Rule*.]

We can also consider $r_{c,p}(a) = \mathbf{P}_a(T_0 < T_c) = \mathbf{P}_a(\text{ruin})$.
Then $r_{c,p}(a)$ is like $s(a)$, but from the other player's perspective.
Hence, we need to replace a by $c - a$, and replace p by $1 - p$.
It follows that:

$$(\mathbf{1.7.3}) \quad r_{c,p}(a) \; = \; s_{c,1-p}(c - a) \; = \; \begin{cases} \dfrac{\left(\frac{p}{1-p}\right)^{c-a} - 1}{\left(\frac{p}{1-p}\right)^{c} - 1}, & p \neq 1/2 \\ (c - a)/c, & p = 1/2 \end{cases}$$

(1.7.4) Problem. Check directly that $r_{c,p}(a) + s_{c,p}(a) = 1$, thus
showing that Gambler's Ruin must eventually end.

Next we consider $\mathbf{P}_a(T_0 < \infty)$, the probability of <u>eventual</u> ruin.
Clearly $\lim_{c \to \infty} T_c = \infty$ (since e.g. $T_c \geq c - a$).
Hence, by continuity of probabilities (A.8.3) and then (1.7.3), we can
compute the probability of eventual ruin as:

$$\mathbf{P}_a(T_0 < \infty) \; = \; \lim_{K \to \infty} \mathbf{P}_a(T_0 < K)$$

$$= \; \lim_{c \to \infty} \mathbf{P}_a(T_0 < T_c) \; = \; \lim_{c \to \infty} r_{c,p}(a)$$

$$(\mathbf{1.7.5}) \qquad = \begin{cases} \frac{0-1}{0-1} = 1, & p < 1/2 \\ \frac{1}{1} = 1, & p = 1/2 \\ \left(\frac{p}{1-p}\right)^{-a}, & p > 1/2 \end{cases}$$

So, eventual ruin is certain if $p \leq 1/2$, but not if $p > 1/2$.

For example, suppose $p = 2/3$ and $a = 2$.
Then $\mathbf{P}(T_0 < \infty) = \left(\frac{2/3}{1-(2/3)}\right)^{-2} = 2^{-2} = 1/4$.
Hence, $\mathbf{P}(T_0 = \infty) = 1 - \mathbf{P}(T_0 < \infty) = 3/4$.
That is, if we start with \$2, and have probability 2/3 of winning each
bet, then we have probability 3/4 of <u>never</u> going broke.

Finally, we consider the time $T = \min(T_0, T_c)$ when the Gambler's
Ruin game ends.

It follows from Problem (1.7.4) that the game must end eventually, with probability 1.

But a simple bound on the tail probabilities of T proves more directly that T is finite and has finite expectation:

(1.7.6) Proposition. Let $T = \min(T_0, T_c)$ be the time when the Gambler's Ruin game ends. Then $\mathbf{P}(T > mc) \leq (1 - p^c)^m$, and $\mathbf{P}(T = \infty) = 0$, and $\mathbf{E}(T) < \infty$.

Proof: If the player ever wins c bets in a row, then the game must be over (since if they haven't already lost, then they will have reached c).

But if $T > mc$, then the player has failed to win c bets in a row, despite having m independent attempts to do so.

Now, the probability of winning c bets in a row is p^c.

Hence, the probability of <u>failing</u> to win c bets in a row is $1 - p^c$.

So, the probability of failing on m independent attemps is $(1 - p^c)^m$.

Hence, $\mathbf{P}(T > mc) \leq (1 - p^c)^m$, as claimed.

Then, by continuity of probabilities (A.8.3),

$$\mathbf{P}(T = \infty) \;=\; \lim_{m \to \infty} \mathbf{P}(T > mc) \;\leq\; \lim_{m \to \infty} (1 - p^c)^m \;=\; 0\,.$$

Finally, using the trick (A.2.1), and a geometric series (A.4.1),

$$\mathbf{E}(T) \;=\; \sum_{i=1}^{\infty} \mathbf{P}(T \geq i) \;\leq\; \sum_{i=0}^{\infty} \mathbf{P}(T \geq i)$$

$$=\; \mathbf{P}(T \geq 0) + \mathbf{P}(T \geq 1) + \mathbf{P}(T \geq 2) + \mathbf{P}(T \geq 3) + \mathbf{P}(T \geq 4) + \ldots$$

$$\leq\; \mathbf{P}(T \geq 0) + \mathbf{P}(T \geq 0) + \ldots + \mathbf{P}(T \geq 0) + \mathbf{P}(T \geq c) + \ldots$$

$$=\; \sum_{j=0}^{\infty} c\, \mathbf{P}(T \geq cj) \;\leq\; \sum_{j=0}^{\infty} c\, (1 - p^c)^j$$

$$=\; \frac{c}{1 - (1 - p^c)} \;=\; \frac{c}{p^c} \;<\; \infty\,. \qquad \blacksquare$$

That is, with probability 1 the Gambler's Ruin game must eventually end, and the time it takes to end has finite expected value.

(1.7.7) Remark. A *betting strategy* (or, *gambling strategy*) involves wagering different amounts on different bets. One common example is the *double 'til you win* strategy (also sometimes called a *martingale* strategy, though it is different from the martingales discussed in the next chapter): First you wager $1, then if you lose you wager $2, then if you lose again you wager $4, etc. As soon as you win any one bet you stop, with total net gain always equal to $1. For any $p > 0$, with probability 1 you will <u>eventually</u> win a bet, so this appears to guarantee a $1 profit. (And then by repeating the strategy, or scaling up the wagers, you can guarantee larger profits too.) What's the catch? Well, you might have a run of very bad luck, and reach your credit limit, and have to stop with a very negative net gain. (For example, if you happen to lose your first 30 bets, then you will be down over a billion dollars, with little hope of recovery!) Indeed, it can be proven that provided there is <u>some</u> finite limit to how much you can lose, then if $p \leq 1/2$ then your expected net gain can never be positive; see e.g. Section 7.3 of Rosenthal (2006), or Dubins and Savage (2014). See also Example (3.3.7) below.

(1.7.8) Problem. Consider s.r.w. with $p = 3/5$, $X_0 = 0$, and $T = \inf\{n \geq 1 : X_n = -2\}$. Compute $\mathbf{P}(T = \infty)$. [sol]

(1.7.9) Problem. Suppose a gambler starts with $7, and repeatedly bets $1, with (independent) probability 2/5 of winning each bet. Compute the probability that:
(a) the gambler will reach $10 before losing all their money.
(b) the gambler will reach $20 before losing all their money.
(c) the gambler will <u>never</u> lose all their money.

(1.7.10) Problem. Suppose $p = 0.49$ and $c = 10,000$. Find the smallest value of a such that $s_{c,p}(a) \geq 1/2$. Interpret your answer in terms of a gambler at a casino.

(1.7.11) Problem. Suppose $a = 9,700$ and $c = 10,000$. Find the smallest value of p such that $s_{c,p}(a) \geq 1/2$. Interpret your answer in terms of a gambler at a casino.

(1.7.12) Problem. Consider the Markov chain with state space $S = \{1, 2, 3, 4\}$, $X_0 = 3$, and transition probabilities specified by

$p_{11} = p_{22} = 1$, $p_{31} = p_{32} = p_{34} = 1/3$, $p_{43} = 3/4$, $p_{41} = 1/6$, and $p_{42} = 1/12$. Compute $\mathbf{P}(T_1 < T_2)$. [Hint: You may wish to set $s(a) = \mathbf{P}_a(T_1 < T_2)$.]

(1.7.13) Problem. Consider the Markov chain with state space $S = \{1, 2, 3, 4\}$, $X_0 = 2$, and transition probabilities specified by $p_{11} = p_{44} = 1$, $p_{21} = 2/3$, $p_{23} = p_{24} = 1/6$, and $p_{32} = p_{34} = 1/2$. Compute $\mathbf{P}(T_1 < T_4)$.

(1.7.14) Problem. Consider the Markov chain with state space $S = \{1, 2, 3, 4\}$, $X_0 = 3$, and transition probabilities given by $p_{11} = p_{22} = 1$, $p_{31} = p_{32} = p_{34} = 1/3$, $p_{41} = 1/3$, $p_{42} = 1/6$, $p_{43} = 1/2$, and $p_{ij} = 0$ otherwise. Let $T = \inf\{n \geq 0 : X_n = 1 \text{ or } 2\}$. Compute $\mathbf{P}(X_T = 1)$.

2. Markov Chain Convergence

Now that we know the basics about Markov chains, we turn to questions about its long-run probabilities:

What happens to $\mathbf{P}(X_n = j)$ for large n?

Does $\lim_{n \to \infty} \mathbf{P}(X_n = j)$ exist? What does it equal?

One clue is given by the following.

Suppose $q_j := \lim_{n \to \infty} \mathbf{P}(X_n = j)$ exists for all $j \in S$.

That is, $\lim_{n \to \infty} \mu_j^{(n)} = q_j$.

Then also $\lim_{n \to \infty} \mu_j^{(n+1)} = q_j$, i.e. $\lim_{n \to \infty} \sum_{i \in S} \mu_i^{(n)} p_{ij} = q_j$.

Assuming we can exchange the limit and sum, $\sum_{i \in S} \left(\lim_{n \to \infty} \mu_i^{(n)} \right) p_{ij} = q_j$, i.e. $\sum_{i \in S} (q_i) p_{ij} = q_j$, i.e. $\sum_{i \in S} q_i p_{ij} = q_j$.

That suggests that the limits $\{q_j\}$, if they exist, must satisfy a certain "stationary" property, as we now discuss.

2.1. Stationary Distributions

A key concept is the following.

(2.1.1) Definition. If π is a *probability distribution* on S (i.e., $\pi_i \geq 0$ for all $i \in S$, and $\sum_{i \in S} \pi_i = 1$), then π is *stationary* for a Markov chain with transition probabilities (p_{ij}) if $\sum_{i \in S} \pi_i p_{ij} = \pi_j$ for all $j \in S$.

In matrix notation: $\pi P = \pi$. (Equivalently: π is a *left eigenvector* for the matrix P with *eigenvalue* 1, cf. (A.12.2).)

Intuitively, if the chain starts with probabilities $\{\pi_i\}$, then it will keep the same probabilities one time unit later.

That is, if $\mu^{(0)} = \pi$, i.e. $\mathbf{P}(X_0 = i) = \pi_i$ for all i, then also $\mu^{(1)} = \nu P = \pi P = \pi$, i.e. $\mu^{(1)} = \mu^{(0)}$.

Similarly, if $\mu^{(n)} = \pi$, i.e. $\mathbf{P}(X_n = i) = \pi_i$ for all i, then $\mu^{(n+1)} = \mu^{(n)} P = \pi P = \pi$, i.e. $\mu^{(n+1)} \mu^{(n)}$.

Then, by induction, $\mu^{(m)} = \pi$ for all $m \geq n$, too. ("stationary")

In particular, if $\nu = \pi$, then $\mu^{(n)} = \pi$ for all $n \in \mathbf{N}$.

Hence, also $\pi P^n = \pi$ for $n = 0, 1, 2, \ldots$, i.e. $\sum_{i \in S} \pi_i p_{ij}^{(n)} = \pi_j$.

(This also follows since e.g. $\pi P^{(2)} = (\pi P) P = \pi P = \pi$.)

In the Frog Walk:

Let π be the *uniform distribution* (A.3.1) on S, i.e. $\pi_i = \frac{1}{20}$ for all $i \in S$.

Then $\pi_i \geq 0$ and $\sum_i \pi_i = 1$.

And, if e.g. $j = 8$, then $\sum_{i \in S} \pi_i p_{i8} = \pi_7 p_{78} + \pi_8 p_{88} + \pi_9 p_{98} = \frac{1}{20}(\frac{1}{3}) + \frac{1}{20}(\frac{1}{3}) + \frac{1}{20}(\frac{1}{3}) = \frac{1}{20} = \pi_j$.

Similarly, for any $j \in S$, $\sum_{i \in S} \pi_i p_{ij} = \frac{1}{20}(\frac{1}{3}) + \frac{1}{20}(\frac{1}{3}) + \frac{1}{20}(\frac{1}{3}) = \pi_j$.

So, $\{\pi_i\}$ is a stationary distribution!

Hence, if $\nu_i = 1/20$ for all i, then $\mathbf{P}(X_1 = i) = 1/20$ for all i, and indeed $\mathbf{P}(X_n = i) = 1/20$ for all i and all $n \geq 1$, too.

More generally, suppose that $|S| < \infty$, and we have a chain which is *doubly stochastic*, i.e. $\sum_{i \in S} p_{ij} = 1$ for all $j \in S$ (in addition to the usual condition that $\sum_{j \in S} p_{ij} = 1$ for all $i \in S$).

(This holds for the Frog Walk.)

Let π be the *uniform distribution* (A.3.1) on S, i.e. $\pi_i = \frac{1}{|S|}$ for all $i \in S$.

Then for all $j \in S$, $\sum_{i \in S} \pi_i p_{ij} = \frac{1}{|S|} \sum_{i \in S} p_{ij} = \frac{1}{|S|}(1) = \frac{1}{|S|} = \pi_j$.

So, $\{\pi_i\}$ is stationary.

Consider the Ehrenfest's Urn example (where $S = \{0, 1, 2, \ldots, d\}$, $p_{ij} = i/d$ for $j = i - 1$, and $p_{ij} = (d - i)/d$ for $j = i + 1$).

Is the uniform distribution $\pi_i = \frac{1}{d+1}$ a stationary distribution? No!

e.g. if $j = 1$, $\sum_{i \in S} \pi_i p_{ij} = \frac{1}{d+1}(p_{01} + p_{21}) = \frac{1}{d+1}(1 + \frac{2}{d}) \neq \frac{1}{d+1} = \pi_j$.

So, we should <u>not</u> take $\pi_i = \frac{1}{d+1}$ for all i.

So, how should we choose π?

(2.1.2) Problem. Consider a Markov chain with state space $S = \{1, 2, 3\}$, and transition probabilities $p_{12} = 1/2$, $p_{13} = 1/2$, $p_{21} = 1/3$, $p_{23} = 2/3$, $p_{31} = 2/5$, $p_{32} = 3/5$, otherwise $p_{ij} = 0$.

(a) Draw a diagram of this Markov chain. [sol]

(b) Compute $p_{11}^{(2)}$. [sol]

(c) Find a distribution π which is stationary for this chain. [sol]

(2.1.3) Problem. Consider a Markov chain with state space $S = \{1, 2, 3\}$ and transition probability matrix

$$P = \begin{pmatrix} 1/4 & 3/4 & 0 \\ 2/3 & 0 & 1/3 \\ 0 & 1/5 & 4/5 \end{pmatrix}.$$

(a) Draw a diagram of this chain.

(b) Find a stationary distribution for this chain.

2.2. Searching for Stationarity

How can we find stationary distributions, or decide if they even exist? One trick which sometimes helps is:

(2.2.1) Definition. A Markov chain is *reversible* (or *time reversible*, or satisfies *detailed balance*) with respect to a probability distribution $\{\pi_i\}$ if $\pi_i p_{ij} = \pi_j p_{ji}$ for all $i, j \in S$.

The usefulness of reversibility is:

(2.2.2) Proposition. If a chain is reversible with respect to π, then π is a stationary distribution.

Proof: Reversibility means $\pi_i p_{ij} = \pi_j p_{ji}$, so then for $j \in S$, $\sum_{i \in S} \pi_i p_{ij} = \sum_{i \in S} \pi_j p_{ji} = \pi_j \sum_{i \in S} p_{ji} = \pi_j(1) = \pi_j.$ ∎

(2.2.3) Problem. Consider a Markov chain with state space $S = \{1, 2, 3, 4\}$, and transition probabilities

$$P = \begin{pmatrix} 2/3 & 1/3 & 0 & 0 \\ 2/3 & 0 & 1/3 & 0 \\ 0 & 2/3 & 0 & 1/3 \\ 0 & 0 & 2/3 & 1/3 \end{pmatrix}.$$

(a) Find a probability distribution π with respect to which this chain is reversible. **[sol]**

(b) Prove that this π is stationary for the chain. **[sol]**

However, the converse to Proposition (2.2.2) is false; it is possible for a chain to have a stationary distribution even if it is <u>not</u> reversible:

(2.2.4) Problem. Let $S = \{1, 2, 3\}$, with transition probabilities $p_{12} = p_{23} = p_{31} = 1$, and let $\pi_1 = \pi_2 = \pi_3 = 1/3$.

(a) Show that π is a stationary distribution for P. [Hint: Either check directly, or use the doubly stochastic property, or find the left eigenvector with eigenvalue 1.]
(b) Show that the chain is <u>not</u> reversible w.r.t. π.

In the Frog Walk, let $\pi_i = 1/20$ for all i.
If $|j - i| \le 1$ or $|j - i| = 19$, then $\pi_i p_{ij} = (1/20)(1/3) = \pi_j p_{ji}$.
Otherwise, both sides equal 0.
So, the chain is reversible with respect to π!
(This provides an easier way to check stationarity.)

What about Ehrenfest's Urn? What π is stationary?
New idea: perhaps each <u>ball</u> is equally likely to be in either Urn.
Then, the number of balls in Urn #1 follows a Binomial$(d, 1/2)$ distribution (A.3.2).
That is, let $\pi_i = \binom{d}{i}(1/2)^i(1/2)^{d-i} = 2^{-d}\binom{d}{i} = 2^{-d}\frac{d!}{i!(d-i)!}$.
Then $\pi_i \ge 0$ and $\sum_i \pi_i = 1$, so π is a distribution.
Is π stationary? We need to check $\sum_{i \in S} \pi_i p_{ij} = \pi_j$ for all $j \in S$.
This can be checked directly, but the calculations are pretty messy; see Problem (2.2.5).

What to do instead? Use reversibility!
We need to check if $\pi_i p_{ij} = \pi_j p_{ji}$ for all $i, j \in S$.
Clearly, both sides are 0 unless $j = i + 1$ or $j = i - 1$.
If $j = i + 1$, then

$$\pi_i p_{ij} = 2^{-d}\binom{d}{i}\frac{d-i}{d} = 2^{-d}\frac{d!}{i!(d-i)!}\frac{d-i}{d} = 2^{-d}\frac{(d-1)!}{i!(d-i-1)!}.$$

Also

$$\pi_j p_{ji} = 2^{-d}\binom{d}{j}\frac{j}{d} = 2^{-d}\frac{d!}{j!(d-j)!}\frac{j}{d}$$

$$= 2^{-d}\frac{(d-1)!}{(j-1)!(d-j)!} = 2^{-d}\frac{(d-1)!}{i!(d-i-1)!} = \pi_i p_{ij}.$$

If $j = i - 1$, then again $\pi_i p_{ij} = \pi_j p_{ji}$ (by a similar calculation, or simply by exchanging i and j in the above).
So, it is reversible. So, π is a stationary distribution!
Intuitively, π_i is larger when i is close to $d/2$, which makes sense.

But does $\mu_i^{(n)} \to \pi_i$? We'll see!

(2.2.5) Problem. (*) Prove by direct calculation that the above π is stationary for Ehrenfest's Urn. [Hint: Don't forget the *Pascal's Triangle* identity that $\binom{d-1}{j-1} + \binom{d-1}{j} = \binom{d}{j}$.]

Next, does simple random walk have a stationary distribution? If $p = 1/2$, then s.s.r.w. is symmetric, so it would be reversible with respect to a "uniform" distribution on $\mathbf{Z}$, but what is that? Or perhaps it doesn't have a stationary distribution at all? How could we prove that? One method is given by:

(2.2.6) Vanishing Probabilities Proposition. If a Markov chain's transition probabilities have $\lim_{n\to\infty} p_{ij}^{(n)} = 0$ for all $i, j \in S$, then the chain does <u>not</u> have a stationary distribution.

Proof: If there were a stationary distribution π, then we would have $\pi_j = \sum_{i \in S} \pi_i p_{ij}^{(n)}$ for any n, so also

$$\pi_j = \lim_{n\to\infty} \pi_j = \lim_{n\to\infty} \sum_{i \in S} \pi_i p_{ij}^{(n)}.$$

Next, we claim that we can exchange the sum and limit.
Indeed, exchanging the limit and sum is justified by the M-test (A.11.1), since if $x_{ni} = \pi_i p_{ij}^{(n)}$, then

$$\sum_{i=1}^{\infty} \sup_n |x_{ni}| = \sum_{i=1}^{\infty} \sup_n |\pi_i p_{ij}^{(n)}| \leq \sum_{i=1}^{\infty} |\pi_i| = 1 < \infty.$$

Hence, $\pi_j = \lim_{n\to\infty} \sum_{i \in S} \pi_i p_{ij}^{(n)} = \sum_{i \in S} \lim_{n\to\infty} \pi_i p_{ij}^{(n)} = \sum_{i \in S} (0) = 0$.
So, we have $\pi_j = 0$ for all j.
But this means that $\sum_j \pi_j = 0$, which is a contraction.
Therefore, there is no stationary distribution. ∎

So can we apply this to simple random walk (s.r.w.)?
Well, $p_{00}^{(n)} = 0$ for n odd.
And, we showed (1.5.7) that $p_{00}^{(n)} \approx [4p(1-p)]^{n/2}\sqrt{2/\pi n}$ for large even n.

But we always have $[4p(1 - p)]^{n/2} \leq 1$.

So, for all large enough n, $p_{00}^{(n)} \leq 2\sqrt{2/\pi n}$ (say).

In particular, for s.r.w. with any value of p:

(2.2.7)
$$\lim_{n \to \infty} p_{00}^{(n)} = 0.$$

Does this mean that $\lim_{n \to \infty} p_{ij}^{(n)} = 0$ for <u>all</u> $i, j \in S$? Yes!

(2.2.8) Vanishing Lemma.   If a Markov chain has <u>some</u> $k, \ell \in S$ with $\lim_{n \to \infty} p_{k\ell}^{(n)} = 0$, then for <u>any</u> $i, j \in S$ with $k \to i$ and $j \to \ell$, $\lim_{n \to \infty} p_{ij}^{(n)} = 0$.

Proof: By (1.6.2), we can find $r, s \in \mathbf{N}$ with $p_{ki}^{(r)} > 0$ and $p_{j\ell}^{(s)} > 0$.

By Chapman-Kolmogorov (1.4.5), $p_{k\ell}^{(r+n+s)} \geq p_{ki}^{(r)} p_{ij}^{(n)} p_{j\ell}^{(s)}$.

Hence, $p_{ij}^{(n)} \leq p_{k\ell}^{(r+n+s)} / (p_{ki}^{(r)} p_{j\ell}^{(s)})$.

But the assumption implies that $\lim_{n \to \infty} \left[p_{k\ell}^{(r+n+s)} / (p_{ki}^{(r)} p_{j\ell}^{(s)}) \right] = 0$.

Also, we always have $p_{ij}^{(n)} \geq 0$.

So, $p_{ij}^{(n)}$ is "sandwiched" between 0 and a sequence converging to 0.

Hence, by the *Sandwich Theorem* (A.4.8), $\lim_{n \to \infty} p_{ij}^{(n)} = 0$.   ∎

This immediately implies:

(2.2.9) Vanishing Together Corollary. For an <u>irreducible</u> Markov chain, either (i) $\lim_{n \to \infty} p_{ij}^{(n)} = 0$ for all $i, j \in S$, or (ii) $\lim_{n \to \infty} p_{ij}^{(n)} \neq 0$ for all $i, j \in S$.

Then, combining this Vanishing Together Corollary with the Vanishing Probabilities Proposition gives:

(2.2.10) Vanishing Probabilities Corollary. If an irreducible Markov chain has $\lim_{n \to \infty} p_{k\ell}^{(n)} = 0$ for <u>some</u> $k, \ell \in S$, then the chain does <u>not</u> have a stationary distribution.

Returning to simple random walk: It is irreducible.

It may be recurrent or transient, depending on p.

However, we know that $\lim_{n \to \infty} p_{00}^{(n)} = 0$.

Hence, by the Vanishing Together Corollary, $p_{ij}^{(n)} \to 0$ for <u>all</u> $i, j \in S$. Then, by the Vanishing Probabilities Corollary, simple random walk does <u>not</u> have a stationary distribution.

So, in particular, simple <u>symmetric</u> random walk (with $p = 1/2$) is in case (a) of the Cases Theorem, but in case (i) of the Vanishing Together Corollary. Hence, for all $i, j \in S$, $\sum_{n=1}^{\infty} p_{ij}^{(n)} = \infty$ even though $\lim_{n \to \infty} p_{ij}^{(n)} = 0$, consistent with (A.4.4).

<u>Aside:</u> We always have $\sum_j p_{ij}^{(n)} = 1$ for all n, so $\lim_{n \to \infty} \sum_j p_{ij}^{(n)} = 1$, even though for simple random walk $\sum_j \lim_{n \to \infty} p_{ij}^{(n)} = 0$. How can this be? It must be that we cannot interchange the sum and limit. In particular, the M-test (A.11.1) does not apply, so the M-test conditions are <u>not</u> satisfied, i.e. we must have $\sum_j \sup_n p_{ij}^{(n)} = \infty$.

It also follows that:

(2.2.11) Transient Not Stationary Corollary. An irreducible and <u>transient</u> Markov chain cannot have a stationary distribution.

Proof: If a chain is irreducible and transient, then by the Transience Equivalences Theorem (1.6.13), $\sum_{n=1}^{\infty} p_{ij}^{(n)} < \infty$ for all $i, j \in S$. Hence, by (A.4.3), also $\lim_{n \to \infty} p_{ij}^{(n)} = 0$ for all $i, j \in S$. Thus, by the Vanishing Probabilities Proposition (2.2.6), there is no stationary distribution.   ∎

If S is <u>infinite</u>, can there be a stationary distribution? Yes!

(2.2.12) Example. Let $S = \mathbf{N} = \{1, 2, 3, \ldots\}$, with $p_{1,1} = 3/4$ and $p_{1,2} = 1/4$, and for $i \geq 2$, $p_{i,i} = p_{i,i+1} = 1/4$ and $p_{i,i-1} = 1/2$.
This chain is clearly irreducible.
But does it have a stationary distribution?
Let $\pi_i = 2^{-i}$ for all $i \in S$, so $\pi_i \geq 0$ and $\sum_i \pi_i = 1$.
Then for any $i \in S$, $\pi_i p_{i,i+1} = 2^{-i}(1/4) = 2^{-i-2}$.
Also, $\pi_{i+1} p_{i+1,i} = 2^{-(i+1)}(1/2) = 2^{-i-2}$. Equal!
And, $\pi_i p_{i,j} = 0$ if $|j - i| \geq 2$.
So, the chain is reversible with respect to π.
So, π is a stationary distribution.
Does this mean $\lim_{n \to \infty} P(X_n = j) = \pi_j = 2^{-j}$ for all $j \in S$? We'll see!

2.3. Obstacles to Convergence

If chain has stationary distribution $\{\pi_i\}$, does $\lim_{n\to\infty} \mathbf{P}[X_n = j] = \pi_j$ for all $j \in S$? Not necessarily!

(2.3.1) Example. Let $S = \{1, 2\}$, $\nu_1 = 1$, and $(p_{ij}) = \begin{pmatrix} 1 & 0 \\ 0 & 1 \end{pmatrix}$.

Figure 8: A diagram of Example (2.3.1).

Let $\pi_1 = \pi_2 = \frac{1}{2}$. Then $\{\pi_i\}$ is stationary. (Check this either directly, or by the doubly stochastic property, or by reversibility.)
But $\lim_{n\to\infty} \mathbf{P}(X_n = 1) = \lim_{n\to\infty} 1 = 1 \neq \frac{1}{2} = \pi_1$.
That is, the chain does <u>not</u> converge to stationarity.
However, this chain is not irreducible (i.e. it is *reducible*).

(2.3.2) Example. Let $S = \{1, 2\}$, $\nu_1 = 1$, and $(p_{ij}) = \begin{pmatrix} 0 & 1 \\ 1 & 0 \end{pmatrix}$.

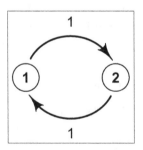

Figure 9: A diagram of Example (2.3.2).

This example is irreducible.
And again, if $\pi_1 = \pi_2 = \frac{1}{2}$, then π is stationary (check!).
But $\mathbf{P}(X_n = 1) = \begin{cases} 1, & n \text{ even} \\ 0, & n \text{ odd} \end{cases}$
So, $\lim_{n\to\infty} \mathbf{P}[X_n = 1]$ does not even exist!
In fact, this chain is "periodic", as we now discuss.

(2.3.3) Definition. The *period* of a state i is the *greatest common divisor (gcd)* of the set $\{n \geq 1 : p_{ii}^{(n)} > 0\}$, i.e. the largest number m such that all values of n with $p_{ii}^{(n)} > 0$ are integer multiples of m. If the period of <u>every</u> state is 1, we say the chain is *aperiodic*. Otherwise we say the chain is *periodic*.

(<u>Note:</u> We assume $f_{ii} > 0$, so that $\{n \geq 1 : p_{ii}^{(n)} > 0\}$ is non-empty.)

For example, if the period of i is 2, this means that it is only possible to get from i to i in an <u>even</u> number of steps. (Like Example (2.3.2) above.)

It follows from the definition of "common divisor" that:

(2.3.4) If state i has period t, and $p_{ii}^{(m)} > 0$, then m is an integer multiple of t, i.e. t divides m.

(2.3.5) Problem.   Let $S = \{1, 2, 3\}$, and $p_{12} = p_{23} = p_{31} = 1$. Show that the period of each state is 3.

(2.3.6) Example. Let $S = \{1, 2, 3\}$, and $(p_{ij}) = \begin{pmatrix} 0 & 1/2 & 1/2 \\ 1/2 & 0 & 1/2 \\ 1/2 & 1/2 & 0 \end{pmatrix}$.

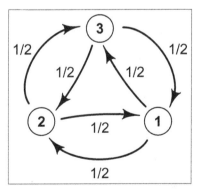

Figure 10: A diagram of Example (2.3.6).

Then it is possible to go from state 1 to state 1 by $1 \rightarrow 2 \rightarrow 1$, or by $1 \rightarrow 2 \rightarrow 3 \rightarrow 1$, or other ways too.

So, the period of state 1 is $gcd\{2, 3, \ldots\} = 1$.

Similarly, all states have period 1.

So, the chain is aperiodic, even though $p_{ii} = 0$ for all $i \in S$.

Now, the gcd of any collection of positive integers which includes 1 must equal 1, i.e. $gcd\{1, \ldots\} = 1$. Hence:

(2.3.7) If $p_{ii} > 0$, then the period of state i is 1.

(The converse to (2.3.7) is false, as shown by Example (2.3.6).)

Also, $gcd\{n, n+1, \ldots\} = 1$, so we have:

(2.3.8) If $p_{ii}^{(n)} > 0$ and $p_{ii}^{(n+1)} > 0$ for some n, then the period of state i is 1.

What about the Frog Walk? There, $p_{ii} = 1/3 > 0$, so each state has period 1, i.e. the chain is aperiodic.

What about Simple Random Walk? It can only return to i after an <u>even</u> number of steps, so the period of each state is 2.

What about Ehrenfest's Urn? Again, it can only return to a state after an <u>even</u> number of steps, so the period of each state is 2.

(2.3.9) Problem. Consider the example from Problem (1.6.19), where $S = \{1, 2, 3\}$, and $p_{11} = p_{12} = p_{22} = p_{23} = p_{32} = p_{33} = 1/2$, with $p_{ij} = 0$ otherwise. Is this chain aperiodic?   [**sol**]

Finally, another helpful property of periodicity is:

(2.3.10) Equal Periods Lemma.   If $i \leftrightarrow j$, then the periods of i and of j are equal.

Proof: Let the periods of i and j be t_i and t_j, respectively. By (1.6.2), find $r, s \in \mathbf{N}$ with $p_{ij}^{(r)} > 0$ and $p_{ji}^{(s)} > 0$.
Then by (1.4.5), $p_{ii}^{(r+s)} \geq p_{ij}^{(r)} p_{ji}^{(s)} > 0$. So, by (2.3.4), t_i divides $r + s$.
Suppose now that $p_{jj}^{(n)} > 0$.
Then $p_{ii}^{(r+n+s)} \geq p_{ij}^{(r)} p_{jj}^{(n)} p_{ji}^{(s)} > 0$, so t_i divides $r + n + s$.
Since t_i divides $r + n + s$ and also divides $r + s$, therefore t_i must divide n.
Since this is true for any n with $p_{jj}^{(n)} > 0$, it follows that t_i is a common divisor of $\{n \in \mathbf{N} : p_{jj}^{(n)} > 0\}$.
But t_j is the <u>greatest</u> such common divisor. So, $t_j \geq t_i$.

Similarly, $t_i \geq t_j$, so $t_i = t_j$.   ∎

(2.3.11) Equal Periods Corollary. If a chain is <u>irreducible</u>, then
all states have the same period.
(So, we can say that e.g. the whole <u>chain</u> has period 3, or the whole
<u>chain</u> is aperiodic, etc.)

Combining this fact with (2.3.7) shows:

(2.3.12) Corollary. If a chain is irreducible, and $p_{ii} > 0$ for <u>some</u>
state i, then the chain is aperiodic.

(2.3.13) Problem. Consider a Markov chain made up of two inter-
secting cycles of lengths 4 and 5, where $S = \{0, 1, 2, 3, 4, 5, 6, 7, 8, 9\}$,
and $p_{01} = p_{05} = 1/2$, and $p_{12} = p_{23} = p_{34} = p_{40} = 1$, and
$p_{56} = p_{67} = p_{78} = p_{89} = p_{90} = 1$, and $p_{ij} = 0$ otherwise.
(a) Draw a diagram of this Markov chain.
(b) Determine whether or not this chain is irreducible.
(c) Compute the period of the state 0.
(d) Compute the period of each state i.
(e) Determine if this chain is aperiodic.

2.4. Convergence Theorem

Fortunately, the above represent *all* of the obstacles to stationarity. That is, once we have found a stationary distribution, and ruled
out the obstacles of reducibility and periodicity, the probabilities
must converge:

(2.4.1) Markov Chain Convergence Theorem. If a Markov
chain is irreducible, and aperiodic, and has a stationary distribution
$\{\pi_i\}$, then $\lim_{n \to \infty} p_{ij}^{(n)} = \pi_j$ for all $i, j \in S$, and $\lim_{n \to \infty} \mathbf{P}(X_n = j) = \pi_j$ for any initial distribution $\{\nu_i\}$.

To prove this (big) theorem, we need two preliminary results.
The first one connects stationary distributions to recurrence.

(2.4.2) Stationary Recurrence Theorem. If a Markov chain is
irreducible, and has a stationary distribution, then it is <u>recurrent</u>.

Proof: The Transient Not Stationary Corollary (2.2.11) says that a chain cannot be irreducible and transient and have a stationary distribution.

So, if a chain is irreducible and has a stationary distribution, then it cannot be transient, i.e. it must be recurrent. ∎

The second preliminary result uses a fact from number theory to provide a more concrete way to use aperiodicity.

(2.4.3) Proposition. If a state i has $f_{ii} > 0$ and is *aperiodic*, then there is $n_0(i) \in \mathbf{N}$ such that $p_{ii}^{(n)} > 0$ for all $n \geq n_0(i)$.

Proof: Let $A = \{n \geq 1 : p_{ii}^{(n)} > 0\}$.

Then A is non-empty since $f_{ii} > 0$.

And, if $m \in A$ and $n \in A$, then $p_{ii}^{(m)} > 0$ and $p_{ii}^{(n)} > 0$, so by the Chapman-Kolmogorov inequality (1.4.5), $p_{ii}^{(m+n)} \geq p_{ii}^{(m)} p_{ii}^{(n)} > 0$, so $m + n \in A$, which shows that A satisfies *additivity*.

Also, $gcd(A) = 1$ since the state i is aperiodic.

Hence, from the Number Theory Lemma (A.13.4), there is $n_0 \in \mathbf{N}$ such that for all $n \geq n_0$ we have $n \in A$, i.e. $p_{ii}^{(n)} > 0$. ∎

(2.4.4) Corollary. If a chain is irreducible and aperiodic, then for any states $i, j \in S$, there is $n_0(i,j) \in \mathbf{N}$ such that $p_{ij}^{(n)} > 0$ for all $n \geq n_0(i,j)$.

Proof: Find $n_0(i)$ as above, and find $m \in \mathbf{N}$ with $p_{ij}^{(m)} > 0$.

Then let $n_0(i,j) = n_0(i) + m$.

Then if $n \geq n_0(i,j)$, then $n - m \geq n_0(i)$, so $p_{ij}^{(n)} \geq p_{ii}^{(n-m)} p_{ij}^{(m)} > 0$. ∎

This next lemma is the key to the proof.

(2.4.5) Markov Forgetting Lemma. If a Markov chain is irreducible and aperiodic, and has stationary distribution $\{\pi_i\}$, then for all $i, j, k \in S$,

$$\lim_{n \to \infty} \left| p_{ik}^{(n)} - p_{jk}^{(n)} \right| = 0 .$$

(Intuitively, after a long time n, the chain "forgets" whether it started from state i or from state j.)

Proof: The proof uses the *coupling* technique of defining two different copies of the chain.

Specifically, define a new Markov chain $\{(X_n, Y_n)\}_{n=0}^{\infty}$, with state space $\overline{S} = S \times S$, and transition probabilities $\overline{p}_{(ij),(k\ell)} = p_{ik} p_{j\ell}$. Intuitively, the new chain has two coordinates, each of which is an independent copy of the original Markov chain.

Since the two copies are independent, this new chain has a joint stationary distribution given by $\overline{\pi}_{(ij)} = \pi_i \pi_j$ for $i, j \in S$.

And, whenever $n \geq \max[n_0(i, k),\ n_0(j, \ell)]$, $\overline{p}_{(ij),(k\ell)}^{(n)} = p_{ij}^{(n)} p_{k\ell}^{(n)} > 0$.

Hence, this new chain is irreducible (since it eventually has positive transition probabilities from anywhere to anywhere), and is aperiodic by (2.3.8).

So, by the Stationary Recurrence Theorem, the new chain is recurrent.

Next, choose any $i_0 \in S$, and let $\tau = \inf\{n \geq 0 : X_n = Y_n = i_0\}$ be the first time that the new chain hits (i_0, i_0), i.e. that both copies of the chain equal i_0 at the same time.

Since the new chain is irreducible and recurrent, the Recurrence Equivalences Theorem (1.6.12) says that $\overline{f}_{(ij),(i_0 i_0)} = 1$. This means that, starting from (i, j), the new chain must eventually hit (i_0, i_0), so $\mathbf{P}_{(ij)}(\tau < \infty) = 1$.

Then, using the *Law of Total Probability* (A.2.3),

$$(2.4.6) \quad p_{ik}^{(n)} = \mathbf{P}_{(ij)}(X_n = k) = \sum_{m=1}^{\infty} \mathbf{P}_{(ij)}(X_n = k,\ \tau = m)$$

$$= \sum_{m=1}^{n} \mathbf{P}_{(ij)}(X_n = k,\ \tau = m) + \mathbf{P}_{(ij)}(X_n = k,\ \tau > n).$$

Similarly,

$$p_{jk}^{(n)} = \sum_{m=1}^{n} \mathbf{P}_{(ij)}(Y_n = k,\ \tau = m) + \mathbf{P}_{(ij)}(Y_n = k,\ \tau > n).$$

On the other hand, if $n \geq m$, then by conditioning (A.6.2),

$$\mathbf{P}_{(ij)}(X_n = k, \tau = m) = \mathbf{P}_{(ij)}(\tau = m)\, \mathbf{P}_{(ij)}(X_n = k \,|\, \tau = m)$$

$$= \mathbf{P}_{(ij)}(\tau = m)\, \mathbf{P}_{(ij)}(X_n = k \,|\, X_m = Y_m = i_0)$$

$$= \mathbf{P}_{(ij)}(\tau = m)\, \mathbf{P}(X_n = k \,|\, X_m = i_0)$$

$$= \mathbf{P}_{(ij)}(\tau = m)\, p_{i_0,k}^{(n-m)}.$$

Similarly,

$$\mathbf{P}_{(ij)}(Y_n = k, \tau = m) = \mathbf{P}_{(ij)}(\tau = m)\, p_{i_0,k}^{(n-m)}.$$

So, $\mathbf{P}_{(ij)}(X_n = k, \tau = m) = \mathbf{P}_{(ij)}(Y_n = k, \tau = m)$. This is key! (Intuitively, it makes sense, since once $X_m = Y_m = i_0$, then X_n and Y_n have the same probabilities after that.)

Hence, for all $i, j, k \in S$, using (2.4.6) and the triangle inequality (A.13.1) and monotonicity of probabilities (A.2.4),

$$|p_{ik}^{(n)} - p_{jk}^{(n)}|$$
$$= \left| \sum_{m=1}^{n} \mathbf{P}_{(ij)}(X_n = k,\ \tau = m) + \mathbf{P}_{(ij)}(X_n = k,\ \tau > n) \right.$$
$$\left. - \sum_{m=1}^{n} \mathbf{P}_{(ij)}(Y_n = k,\ \tau = m) - \mathbf{P}_{(ij)}(Y_n = k,\ \tau > n) \right|$$
$$= \left| \mathbf{P}_{(ij)}(X_n = k,\ \tau > n) - \mathbf{P}_{(ij)}(Y_n = k,\ \tau > n) \right|$$
$$\leq \left| \mathbf{P}_{(ij)}(X_n = k,\ \tau > n) \right| + \left| \mathbf{P}_{(ij)}(Y_n = k,\ \tau > n) \right|$$
$$\leq \mathbf{P}_{(ij)}(\tau > n) + \mathbf{P}_{(ij)}(\tau > n) = 2\, \mathbf{P}_{(ij)}(\tau > n).$$

Then, using (A.8.3), $\lim_{n \to \infty} 2\, \mathbf{P}_{(ij)}(\tau > n) = 2\, \mathbf{P}_{(ij)}(\tau = \infty) = 0$, since $\mathbf{P}_{(ij)}(\tau < \infty) = 1$. This gives the result. ∎

Aside: The above factor "2" isn't really necessary, since $\mathbf{P}_{(ij)}(X_n = k,\ \tau > n)$ and $\mathbf{P}_{(ij)}(Y_n = k,\ \tau > n)$ are each between 0 and $\mathbf{P}_{(ij)}(\tau > n)$.

Proof of the Markov Chain Convergence Theorem (2.4.1):

Intuition: The Markov Forgetting Lemma says the intial probabilities don't matter in the long run. So, we might as well have started in the stationary distribution. But then we would always remain in the stationary distribution, too!

More precisely, using the triangle inequality (A.13.1),

$$\left| p_{ij}^{(n)} - \pi_j \right| = \left| \sum_{k \in S} \pi_k \left(p_{ij}^{(n)} - p_{kj}^{(n)} \right) \right| \le \sum_{k \in S} \pi_k \left| p_{ij}^{(n)} - p_{kj}^{(n)} \right| .$$

By the Markov Forgetting Lemma (2.4.5), $\lim_{n \to \infty} \pi_k \left| p_{ij}^{(n)} - p_{kj}^{(n)} \right| = 0$ for all $k \in S$.

We next need to exchange the limit and sum.

Indeed, $\sup_n |p_{ij}^{(n)} - p_{kj}^{(n)}| \le 2$ by the triangle inequality (A.13.1), so we have $\sum_{k \in S} \sup_n \pi_k |p_{ij}^{(n)} - p_{kj}^{(n)}| \le \sum_{k \in S} 2\pi_k = 2 < \infty$.

Hence, by the M-test (A.11.1), the exchange is justified.

We conclude that

$$\lim_{n \to \infty} \left| p_{ij}^{(n)} - \pi_j \right| \le \lim_{n \to \infty} \sum_{k \in S} \pi_k \left| p_{ij}^{(n)} - p_{kj}^{(n)} \right|$$

$$= \sum_{k \in S} \lim_{n \to \infty} \pi_k \left| p_{ij}^{(n)} - p_{kj}^{(n)} \right| = \sum_{k \in S} (0) = 0 .$$

Hence, $\lim_{n \to \infty} p_{ij}^{(n)} = \pi_j$ for all $i, j \in S$, as claimed.

Finally, for <u>any</u> $\{\nu_i\}$, we have (with the exchange justified as above):

$$\lim_{n \to \infty} \mathbf{P}(X_n = j) = \lim_{n \to \infty} \sum_{i \in S} \mathbf{P}(X_0 = i, \ X_n = j) = \lim_{n \to \infty} \sum_{i \in S} \nu_i \, p_{ij}^{(n)}$$

$$= \sum_{i \in S} \nu_i \lim_{n \to \infty} p_{ij}^{(n)} = \sum_{i \in S} \nu_i \, \pi_j = (1)\pi_j = \pi_j . \qquad \blacksquare$$

So, for the Frog Walk, regardless of the choice of $\{\nu_i\}$, we have $\lim_{n \to \infty} \mathbf{P}(X_n = 14) = 1/20$, etc.

Convergence follows similarly for all other examples which are irreducible and aperiodic and have a stationary distribution.

We showed that Example (2.2.12) (with $S = \mathbf{N} = \{1, 2, 3, \dots\}$ and $p_{1,1} = 3/4$ and $p_{1,2} = 1/4$, and for $i \ge 2$, $p_{i,i} = p_{i,i+1} = 1/4$ and $p_{i,i-1} = 1/2$), is irreducible and aperiodic with stationary distribution $\pi_i = 2^{-i}$.

Hence, by the Markov Chain Convergence Theorem (2.4.1), this example has $\lim_{n \to \infty} \mathbf{P}(X_n = j) = \pi_j = 2^{-j}$ for all $j \in S$.

Theorem (2.4.1) says nothing about *quantitative convergence rates*, i.e. how <u>large</u> n has to be to make $|\mathbf{P}(X_n = j) - \pi_j| < \epsilon$ for given $\epsilon > 0$. This is an active area of research[1], but it is difficult since it depends on all of the details of all of the Markov chain transitions, and we do not pursue it further here.

Another application of (2.4.1) (generalised in (2.5.3) below) is:

(2.4.7) Corollary. If a Markov chain is irreducible and aperiodic, then it has at most <u>one</u> stationary distribution.

Proof: By Theorem (2.4.1), <u>any</u> stationary distribution must be equal to $\lim_{n\to\infty} \mathbf{P}(X_n = j)$, so they're all equal. ∎

(2.4.8) Example. Let $S = \{1, 2, 3\}$, and $(p_{ij}) = \begin{pmatrix} 2/3 & 1/3 & 0 \\ 1/3 & 2/3 & 0 \\ 0 & 0 & 1 \end{pmatrix}$.

Some stationary distributions for this chain (check!) are:
Stationary distribution #1: $\pi_1 = \pi_2 = \pi_3 = 1/3$ (which follow since the chain is doubly stochastic).
Stationary distribution #2: $\pi_1 = \pi_2 = 1/2$ and $\pi_3 = 0$.
Stationary distribution #3: $\pi_1 = \pi_2 = 0$ and $\pi_3 = 1$.
Stationary distribution #4: $\pi_1 = \pi_2 = 1/8$ and $\pi_3 = 3/4$.
So, in this example the stationary distribution is <u>not</u> unique.
This chain is not <u>irreducible</u>, so Corollary (2.4.7) does not apply.

(2.4.9) Example. Let $S = \{1, 2, 3\}$, and

$$(p_{ij}) = \begin{pmatrix} 0 & 1/2 & 1/2 \\ 1/3 & 1/3 & 1/3 \\ 1/4 & 1/4 & 1/2 \end{pmatrix}.$$

Is it irreducible? Yes! e.g. $f_{11} \geq p_{12}\, p_{21} = (1/2)(1/3) > 0$, etc.
Is it aperiodic? Yes! e.g. $p_{33} = 1/2 > 0$, and irreducible.
Does it have a stationary distribution? If yes, what is it?
Well, we need $\pi P = \pi$, i.e. $\sum_{i \in S} \pi_i p_{ij} = \pi_j$ for all $j \in S$.
$j = 1$: $\pi_1(0) + \pi_2(1/3) + \pi_3(1/4) = \pi_1$.
$j = 2$: $\pi_1(1/2) + \pi_2(1/3) + \pi_3(1/4) = \pi_2$.

---

[1]For an introduction, see e.g. Rosenthal (1995).

$j = 3$: $\pi_1(1/2) + \pi_2(1/3) + \pi_3(1/2) = \pi_3$.
Subtract $j=2$ from $j=3$: $\pi_3[(1/2)-(1/4)] = \pi_3-\pi_2$, i.e. $\pi_2 = \pi_3(3/4)$.
Substitute this into $j=1$: $\pi_1 = \pi_2(1/3) + \pi_3(1/4) = \pi_3(3/4)(1/3) + \pi_3(1/4) = \pi_3/2$.
Then, we need $\pi_1 + \pi_2 + \pi_3 = 1$, i.e. $\pi_3/2 + \pi_3(3/4) + \pi_3 = 1$, i.e. $\pi_3(9/4) = 1$, so we must have $\pi_3 = 4/9$.
Then, $\pi_2 = \pi_3(3/4) = (4/9)(3/4) = 1/3$.
And, $\pi_1 = \pi_3/2 = (4/9)/2 = 2/9$.
Hence, a stationary distribution is $\pi = (2/9, 1/3, 4/9)$.
It can then be checked directly that $\pi P = \pi$.
And, this stationary distribution is unique by (2.4.7).
Finally, does $\lim_{n\to\infty} p_{ij}^{(n)} = \pi_j$ for all $i, j \in S$?
Yes, by the Markov Chain Convergence Theorem (2.4.1)!

Finally, consider convergence of simple random walk (s.r.w.).
We know by (2.2.10) that s.r.w. has no stationary distribution, so it does not converge in the sense of (2.4.1).
But we know (1.6.17) that w.p. 1, if $p > 1/2$ then s.r.w. converges to $+\infty$, or if $p < 1/2$, then s.r.w. converges to $-\infty$.
But what about simple <u>symmetric</u> random walk, with $p = 1/2$?
Well, we know (1.5.8) that s.s.r.w. is recurrent, so $f_{ij} = 1$ for all $i, j \in$ **N**, and it has infinite *fluctuations* (1.6.14), i.e. w.p. 1 it eventually hits every possible collection of integers, in sequence.
So, s.s.r.w. certainly doesn't converge to anything w.p. 1.
Its absolute values $|X_n|$ still have infinite fluctuations on the non-negative integers, so they don't converge to anything w.p. 1 either.
However, the absolute values still converge in some sense:

(2.4.10) Proposition. If $\{X_n\}$ is simple symmetric random walk, then the absolute values $|X_n|$ converge weakly to positive infinity. (So, $|X_n|$ converges weakly but not strongly; cf. Section A.7.)

Proof: We can write (similar to the proof of (1.6.17)) that $X_n = \sum_{i=1}^n Z_i$, where $\{Z_i\}$ are i.i.d. ± 1 with probability $1/2$ each, and hence common mean $m = 0$ and variance $v = 1$.
Hence, for any $b > 0$, the Central Limit Theorem (A.7.3) says that

$$\lim_{n\to\infty} \mathbf{P}\left(-b < \frac{X_n}{\sqrt{n}} < b\right) = \int_{-b}^{b} \phi(x)\, dx,$$

where ϕ is the standard normal distribution's density function as in (A.3.11).

Hence, $\lim_{b \searrow 0} \lim_{n \to \infty} \mathbf{P}(-b < \frac{X_n}{\sqrt{n}} < b) = 0$.

But $\mathbf{P}(|X_n| < K) = \mathbf{P}(-K < X_n < K) = \mathbf{P}(\frac{-K}{\sqrt{n}} < \frac{X_n}{\sqrt{n}} < \frac{K}{\sqrt{n}})$.

So, by monotonicity (A.2.4), for any fixed finite K and any $b > 0$, $\mathbf{P}(|X_n| < K) \leq \mathbf{P}(-b < \frac{X_n}{\sqrt{n}} < b)$ for all $n \geq K^2/b^2$, i.e. all sufficiently large n.

It follows that $\lim_{n \to \infty} \mathbf{P}(|X_n| < K) = 0$ for any fixed finite K.

Thus, as in (A.7.1), $|X_n|$ converges weakly to positive infinity. ∎

(2.4.11) Problem. Consider a Markov chain on the state space $S = \{1, 2, 3, 4\}$ with the following transition matrix:

$$P = \begin{pmatrix} 0.1 & 0.2 & 0.5 & 0.2 \\ 0.4 & 0.3 & 0.2 & 0.1 \\ 0.3 & 0.2 & 0.1 & 0.4 \\ 0.2 & 0.3 & 0.2 & 0.3 \end{pmatrix}$$

Let π be the uniform distribution on S, so $\pi_i = 1/4$ for all $i \in S$.

(a) Compute $p_{14}^{(2)}$. [sol]

(b) Is this Markov chain reversible with respect to π? [sol]

(c) Is π a stationary distribution for this Markov chain? [sol]

(d) Does $\lim_{n \to \infty} p_{ij}^{(n)} = \pi_j$ for all $i, j \in S$? [sol]

(2.4.12) Problem. Consider the Markov chain from Problem (1.4.6), with state space $S = \{1, 2\}$, and transition probabilities $p_{11} = 2/3$, $p_{12} = 1/3$, $p_{21} = 1/4$, and $p_{22} = 3/4$, which is irreducible by Problem (1.6.18).

(a) Determine whether or not this chain is aperiodic. [sol]

(b) Find a stationary distribution $\{\pi_i\}$ for this chain. [sol]

(c) Determine whether or not $\lim_{n \to \infty} p_{12}^{(n)} = \pi_2$. [sol]

(2.4.13) Problem. Suppose there are 10 lily pads arranged in a circle, numbered consecutively clockwise from 1 to 10. A frog begins on lily pad #1. Each second, the frog jumps one pad clockwise with probability 1/4, or two pads clockwise with probability 3/4.

(a) Find a state space S, initial probabilities $\{\nu_i\}$, and transition probabilities $\{p_{ij}\}$, with respect to which this process is a Markov chain. [sol]

(b) Determine if this Markov chain is irreducible. [sol]

(c) Determine if this Markov chain is aperiodic, or if not then what its period equals. [sol]

(d) Determine whether or not $\sum_{n=1}^{\infty} p_{15}^{(n)} = \infty$. [sol]

(e) Either find a stationary distribution $\{\pi_i\}$ for this chain, or prove that no stationary distribution exists. [sol]

(f) Determine if $\lim_{n \to \infty} p_{15}^{(n)}$ exists, and if yes what it equals. [sol]

(2.4.14) Problem. For each of the following sets of conditions, either provide (with explanation) an example of a state space S and Markov chain transition probabilities $\{p_{ij}\}_{i,j \in S}$ such that the conditions are satisfied, or prove that no such Markov chain exists.

(a) The chain is irreducible and periodic (i.e., not aperiodic), and has a stationary probability distribution. [sol]

(b) The chain is irreducible, and there are states $k \in S$ with period 2, and $\ell \in S$ with period 4. [sol]

(c) The chain is irreducible and transient, and is reversible with respect to some probability distribution π. [sol]

(d) The chain is irreducible with stationary distribution π, and $p_{ij} < 1$ for all $i, j \in S$, but the chain is <u>not</u> reversible w.r.t. π. [sol]

(e) There are distinct states $i, j, k \in S$ with $p_{ij} > 0$, $p_{jk}^{(2)} > 0$, and $p_{ki}^{(3)} > 0$, and the state i is periodic (i.e., has period > 1). [sol]

(f) $p_{11} > 1/2$, and the state 1 has period 2.

(g) $p_{11} > 1/2$, and state 2 has period 3, and the chain is irreducible.

(2.4.15) Problem. Let $S = \mathbf{Z}$ (the set of all integers), and let $h : S \to (0, 1)$ with $h(i) > 0$ for all $i \in S$, and $\sum_{i \in S} h(i) = 1$. Consider the transition probabilities on S given by $p_{ij} = (1/4) \min(1, h(j)/h(i))$ if $j = i-2, i-1, i+1$, or $i+2$, and $p_{ii} = 1 - p_{i,i-2} - p_{i,i-1} - p_{i,i+1} - p_{i,i+2}$, and $p_{ij} = 0$ whenever $|j - i| \geq 3$. Prove that $\lim_{n \to \infty} p_{ij}^{(n)} = h(j)$ for all $i, j \in S$. (Hint: Reversibility might help.)

(2.4.16) Problem. For each of the following sets of conditions, either provide (with explanation) an example of a state space S and

Markov chain transition probabilities $\{p_{ij}\}_{i,j \in S}$ such that the conditions are satisfied, or prove that no such Markov chain exists.

(a) The chain is irreducible, with period 3, and has a stationary distribution. **[sol]**

(b) There is $k \in S$ with period 2, and $\ell \in S$ with period 4. **[sol]**

(c) The chain has a stationary distribution π, and $0 < p_{ij} < 1$ for all $i, j \in S$, but the chain is <u>not</u> reversible with respect to π. **[sol]**

(d) The chain is irreducible, and there are distinct states $i, j, k, \ell \in S$ such that $f_{ij} < 1$, and $\sum_{n=1}^{\infty} p_{k\ell}^{(n)} = \infty$. **[sol]**

(e) The chain is irreducible, and there are distinct states $i, j, k \in S$ with $p_{ij} > 0$, $p_{jk}^{(2)} > 0$, and $p_{ki}^{(3)} > 0$, and state i is <u>periodic</u> with period equal to an <u>odd</u> number. **[sol]**

(f) There are states $i, j \in S$ with $0 < f_{ij} < 1$, and $p_{ij}^{(n)} = 0$ for all $n \in \mathbf{N}$. **[sol]**

(g) There are states $i, j \in S$ with $0 < f_{ij} < 1$, and $p_{ij}^{(n)} = 0$ for all $n \geq 2$. **[sol]**

(h) There are distinct states $i, j \in S$ with $f_{ij} > 0$ and $f_{ji} > 0$, and i is transient. **[sol]**

(i) There are distinct states $i, j, k \in S$ with $f_{ij} = 1/2$, $f_{jk} = 1/3$, and $f_{ik} = 1/10$. **[sol]**

(2.4.17) Problem. Consider a Markov chain with state space $S = \{1, 2, 3\}$, and transition probabilities $p_{12} = 1/2$, $p_{13} = 1/2$, $p_{21} = 1/3$, $p_{23} = 2/3$, $p_{31} = 1/4$, $p_{32} = 3/4$.

(a) Compute $p_{11}^{(2)}$.

(b) Compute $p_{13}^{(3)}$.

(c) Find a stationary distribution π for this chain.

(d) Determine (with explanation) whether or not $\lim_{n \to \infty} p_{13}^{(n)} = \pi_3$.

(e) Determine (with explanation) whether or not $f_{13} = 1$.

(f) Determine (with explanation) whether or not $\sum_{n=1}^{\infty} p_{13}^{(n)} = \infty$.

(2.4.18) Problem. Consider a Markov chain on the state space $S = \{1, 2, 3\}$ with transition matrix:

$$P = \begin{pmatrix} 1/3 & 0 & 2/3 \\ 1/6 & 1/3 & 1/2 \\ 4/5 & 0 & 1/5 \end{pmatrix}$$

(a) Compute $p_{13}^{(2)}$.

(b) Specify (with explanation) which states are recurrent, and which states are transient.

(c) Compute f_{23}.

(d) Find a stationary distribution $\{\pi_i\}$ for the chain.

(e) Determine if the chain reversible with respect to π.

(f) Determine whether or not $\lim_{n\to\infty} p_{1j}^{(n)} = \pi_j$ for all $j \in S$. [Hint: Don't forget the Closed Subset Note (1.6.16).]

(2.4.19) Problem. Consider the Markov chain from Problem (2.1.2), with state space $S = \{1, 2, 3\}$, and transition probabilities $p_{12} = 1/2$, $p_{13} = 1/2$, $p_{21} = 1/3$, $p_{23} = 2/3$, $p_{31} = 2/5$, $p_{32} = 3/5$, otherwise $p_{ij} = 0$, and stationary distribution π. Determine whether or not $\lim_{n\to\infty} p_{ij}^{(n)} = \pi_j$ for all $i, j \in S$. [**sol**]

(2.4.20) Problem. Suppose there are 10 lily pads arranged in a circle, numbered consecutively clockwise from 1 to 10. A frog begins on lily pad #1. Each second, the frog jumps one pad clockwise with probability 1/4, or <u>two</u> pads clockwise with probability 3/4.

(a) Find a state space S, initial probabilities $\{\nu_i\}$, and transition probabilities $\{p_{ij}\}$, with respect to which this process is a Markov chain. [**sol**]

(b) Determine if this Markov chain is irreducible. [**sol**]

(c) Determine if this Markov chain is aperiodic, or if not then what its period equals. [**sol**]

(d) Determine whether or not $\sum_{n=1}^{\infty} p_{23}^{(n)} = \infty$. [**sol**]

(e) Either find a stationary distribution $\{\pi_i\}$ for this chain, or prove that no stationary distribution exists. [**sol**]

(f) Determine whether or not $\lim_{n\to\infty} p_{23}^{(n)}$ exists, and if so then what it equals. [**sol**]

(g) Determine whether or not $\lim_{n\to\infty} \frac{1}{2}[p_{23}^{(n)} + p_{23}^{(n+1)}]$ exists, and if so then what it equals. [**sol**]

2.5. Periodic Convergence

Next, we consider periodic chains (e.g. s.r.w., Ehrenfest, etc.).

Recall Example (2.3.2): $S = \{1, 2\}$, $\nu_1 = 1$, $(p_{ij}) = \begin{pmatrix} 0 & 1 \\ 1 & 0 \end{pmatrix}$.
This example is irreducible, and has period $b = 2$, and the uniform distribution $\pi_1 = \pi_2 = \frac{1}{2}$ is stationary.

Here $\mathbf{P}(X_n = 1) = \begin{cases} 1, & n \text{ even} \\ 0, & n \text{ odd} \end{cases}$

Do these transition probabilities converge to a stationary distribution? No! They oscillate between zero and positive values.

However, if we <u>average</u> over different times (i.e. both odd and even times), then the average transition probabilities <u>do</u> converge to π.
In this example, $\lim_{n \to \infty} \frac{1}{2} [p_{ij}^{(n)} + p_{ij}^{(n+1)}] = \pi_j = 1/2$.
Similar convergence holds in general:

(2.5.1) Periodic Convergence Theorem. Suppose a Markov chain is irreducible, with period $b \geq 2$, and stationary distribution $\{\pi_i\}$. Then for all $i, j \in S$,

$$\lim_{n \to \infty} \frac{1}{b} [p_{ij}^{(n)} + \ldots + p_{ij}^{(n+b-1)}] = \pi_j,$$

and

$$\lim_{n \to \infty} \frac{1}{b} \left(\mathbf{P}[X_n = j] + \mathbf{P}[X_{n+1} = j] + \ldots + \mathbf{P}[X_{n+b-1} = j] \right) = \pi_j,$$

and also

$$\lim_{n \to \infty} \frac{1}{b} \mathbf{P}[X_n = j \text{ or } X_{n+1} = j \text{ or } \ldots \text{ or } X_{n+b-1} = j] = \pi_j.$$

For example, if $b = 2$, then $\lim_{n \to \infty} \frac{1}{2} [p_{ij}^{(n)} + p_{ij}^{(n+1)}] = \pi_j$ for all $i, j \in S$.
Of course, in the <u>aperiodic</u> case, by (2.4.1) we also have $\lim_{n \to \infty} \frac{1}{2} [p_{ij}^{(n)} + p_{ij}^{(n+1)}] = \pi_j$ for all $i, j \in S$.

The proof of the Periodic Convergence Theorem is not overly difficult, but it requires a surprisingly large number of steps, so it is deferred to Problems (2.5.6) and (2.5.7) below.

In particular, Problem (2.5.6) first proves the *Cyclic Decomposition Lemma*: there is a disjoint partition $S = S_0 \overset{\bullet}{\cup} S_1 \overset{\bullet}{\cup} \ldots \overset{\bullet}{\cup} S_{b-1}$

such that with probability 1, the chain always moves from S_0 to S_1, and then to S_2, and so on to S_{b-1}, and then back to S_0.

For example, *Ehrenfest's Urn* has period $b = 2$, and oscillates between the two subsets $S_0 = \{$even $i \in S\}$ and $S_1 = \{$odd $i \in S\}$. And, it satisfies the periodic convergence property that

$$\lim_{n \to \infty} \frac{1}{2} [p_{ij}^{(n)} + p_{ij}^{(n+1)}] = \pi_j = 2^{-d} \binom{d}{j}.$$

One type of convergence holds for all irreducible chains with stationary distributions, whether periodic or not:

(2.5.2) Average Probability Convergence. If a Markov chain is irreducible with stationary distribution $\{\pi_i\}$ (whether periodic or not), then for all $i, j \in S$, $\lim_{n \to \infty} \frac{1}{n} [p_{ij}^{(1)} + p_{ij}^{(2)} + \ldots + p_{ij}^{(n)}] = \pi_j$, i.e. $\lim_{n \to \infty} \frac{1}{n} \sum_{\ell=1}^{n} p_{ij}^{(\ell)} = \pi_j$.

Proof: This follows from either the usual Markov Chain Convergence Theorem (for aperiodic chains), or the Periodic Markov Chain Convergence Theorem (for chains with period $b \geq 2$), by the *Cesàro sum* principle (A.4.7): if a sequence converges, then its partial averages also converge to the same value. ∎

One use of this result is to generalise (2.4.7) to periodic chains:

(2.5.3) Unique Stationary Corollary. An irreducible chain, whether periodic or not, has at most <u>one</u> stationary distribution.

Proof: By Average Probability Convergence (2.5.2), any stationary distribution which exists must be equal to $\lim_{n \to \infty} \frac{1}{n} [p_{ij}^{(1)} + p_{ij}^{(2)} + \ldots + p_{ij}^{(n)}]$, so they're all equal. ∎

(2.5.4) Problem. Let $S = \{1, 2, 3\}$, $p_{11} = 1$, and $p_{22} = p_{23} = p_{32} = p_{33} = 1/2$. For $a \in [0, 1]$, let π^a be the probability distribution on S defined by $\pi_1^a = a$, and $\pi_2^a = \pi_3^a = (1 - a)/2$.
(a) Prove that for any $a \in [0, 1]$, π^a is a stationary distribution.
(b) How many stationary distributions does this chain have?

(c) Why does this not contradict the Unique Stationary Corollary?

(2.5.5) Problem. Consider a Markov chain with state space $S = \{1, 2, 3\}$, and transitions $p_{12} = p_{32} = 1$, $p_{21} = 1/4$, and $p_{23} = 3/4$, with $p_{ij} = 0$ otherwise. Let $\pi_1 = 1/8$, $\pi_2 = 1/2$, and $\pi_3 = 3/8$.
(a) Verify that the chain is reversible with respect to π. [sol]
(b) Determine (with explanation) which of the following statements are true and which are false: (i) $\lim_{n \to \infty} p_{11}^{(n)} = 1/8$. (ii) $\lim_{n \to \infty} \frac{1}{2}[p_{11}^{(n)} + p_{11}^{(n+1)}] = 1/8$. (iii) $\lim_{n \to \infty} \frac{1}{n} \sum_{\ell=1}^{n} p_{11}^{(\ell)} = 1/8$. [sol]

(2.5.6) Problem. (*) The *Cyclic Decomposition Lemma* says that for an irreducible Markov chain with period $b \geq 2$, there is a "cyclic" disjoint partition $S = S_0 \dot\cup S_1 \dot\cup \ldots \dot\cup S_{b-1}$ such that if $i \in S_r$ for some $0 \leq r \leq b-2$, then $\sum_{j \in S_{r+1}} p_{ij} = 1$, while if $i \in S_{b-1}$, then $\sum_{j \in S_0} p_{ij} = 1$. (That is, the chain is forced to repeatedly move from S_0 to S_1 to S_2 to $\ldots$ to S_{b-1} and then back to S_0.) Furthermore, the b-step chain $P^{(b)}$, when restricted to S_0, is irreducible and aperiodic. Prove this lemma, as follows:
(a) Fix $i_0 \in S$, and let $S_r = \{j \in S : p_{i_0 j}^{(bm+r)} > 0 \text{ for some } m \in \mathbf{N}\}$. Show that $\bigcup_{r=0}^{b-1} S_r = S$. [Hint: Use irreducibility.]
(b) Show that if $0 \leq r < t \leq b-1$, then S_r and S_t are disjoint, i.e. $S_r \cap S_t = \emptyset$. [Hint: Suppose $j \in S_r \cap S_t$. Find $m \in \mathbf{N}$ with $p_{ji}^{(m)} > 0$. What can you conclude about $\gcd\{n \geq 1 : p_{ii}^{(n)} > 0\}$?]
(c) Let $i \in S_r$ for some $0 \leq r \leq b-2$, and $p_{ij} > 0$. Prove that $j \in S_{r+1}$.
(d) Suppose $i \in S_{b-1}$, and $p_{ij} > 0$. Prove that $j \in S_0$.
(e) Let $\widehat{P} = P^{(b)}|_{S_0}$, i.e. $\widehat{p}_{ij} = p_{ij}^{(b)}$ corresponding to b steps of the original chain, except restricted to just $i, j \in S_0$. Prove that $\widehat{P}$ is irreducible on S_0. [Hint: Suppose there are $i, j \in S_0$ with $\widehat{p}_{ij}^{(n)} = 0$ for all $n \geq 1$. What does this say about f_{ij} for the original chain?]
(f) Prove that $\widehat{P}$ is aperiodic. [Hint: Suppose there is $i \in S_0$ with $\gcd\{n \geq 1 : \widehat{p}_{ii}^{(n)} > 0\} = m \geq 2$. What does this say about the period of i in the original chain?]

(2.5.7) Problem. (*) Complete the proof of the *Periodic Convergence Theorem*, by the following steps. Assume $\{p_{ij}\}$ are the transition probabilities for an irreducible Markov chain on a state space S with period $b \geq 2$, and stationary distribution $\{\pi_i\}$. Let $S_0, S_1, \ldots, S_{b-1}$ and $\widehat{P}$ be as in *Cyclic Decomposition Lemma* in the previous problem. For any $A \subseteq S$, write $\pi(A) = \sum_{i \in A} \pi_i$ for the total probability of A according to π.

(a) Prove that $\pi(S_0) = \pi(S_1) = \ldots = \pi(S_{b-1}) = 1/b$. [Hint: What is the relationship between $\mathbf{P}[X_0 \in S_0]$ and $\mathbf{P}[X_1 \in S_1]$? On the other hand, if we begin in stationarity, then how does $\mathbf{P}[X_n \in S_0]$ change with n?]

(b) Let $\widehat{\pi}_i = b\,\pi_i$ for all $i \in S_0$. Show that $\widehat{\pi}$ is stationary for $\widehat{P}$.

(c) Conclude that $\lim_{m \to \infty} \widehat{p}_{ij}^{(m)} = \widehat{\pi}_j$ for all $i, j \in S_0$, i.e. $\lim_{m \to \infty} p_{ij}^{(bm)} = b\,\pi_j$.

(d) Prove similarly that $\lim_{m \to \infty} p_{ij}^{(bm)} = b\,\pi_j$ for all $i, j \in S_r$ and $1 \leq r \leq b - 1$.

(e) Show that for $i \in S_0$ and $j \in S_r$ for any $1 \leq r \leq b - 1$, we have $\lim_{m \to \infty} p_{ij}^{(bm+r)} = b\,\pi_j$. [Hint: $p_{ij}^{(bm+r)} = \sum_{k \in S} p_{ik}^{(r)} p_{kj}^{(bm)}$.]

(f) Conclude that for $i \in S_0$, $\lim_{m \to \infty} \frac{1}{b}[p_{ij}^{(bm)} + p_{ij}^{(bm+1)} + \ldots + p_{ij}^{(bm+b-1)}] = \pi_j$ for any $j \in S$.

(g) Show that the previous statement holds for any $i \in S$ (not just $i \in S_0$).

(h) Show that $\lim_{n \to \infty} \frac{1}{b}[p_{ij}^{(n)} + \ldots + p_{ij}^{(n+b-1)}] = \pi_j$, i.e. we can take the limit over all n not just $n = bm$. [Hint: For any n, let $m = \lfloor n/b \rfloor$ be the *floor* of n/b, i.e. the greatest integer not exceeding n/b.]

2.6. Application – MCMC Algorithms (Discrete)

Let $S = \mathbf{Z}$, or more generally let S be any continguous subset of $\mathbf{Z}$, e.g. $S = \{1, 2, 3\}$, or $S = \{-5, -4, \ldots, 17\}$, or $S = \mathbf{N}$, etc. Let $\{\pi_i\}$ be any probability distribution on S. Assume for simplicity that $\pi_i > 0$ for all $i \in S$.

Suppose we want to *sample* from π, i.e. create random variables X with $\mathbf{P}(X = i) \approx \pi_i$ for all $i \in S$.

(This approach is called *Monte Carlo*, and allows us to estimate probabilities and expected values w.r.t. π.)

One popular method is *Markov chain Monte Carlo (MCMC)*:

Create a <u>Markov chain</u> $X_0, X_1, X_2, \ldots$ such that $\lim_{n \to \infty} \mathbf{P}(X_n = i) = \pi_i$ for all $i \in S$, so for large n, $\mathbf{P}(X_n = i) \approx \pi_i$ for all $i \in S$. Can we do this? Can we <u>create</u> Markov chain transitions $\{p_{ij}\}$ so that $\lim_{n \to \infty} p_{ij}^{(n)} = \pi_j$?

Yes! One way is the *Metropolis algorithm*[2]:
Let $p_{i,i+1} = \frac{1}{2} \min[1, \frac{\pi_{i+1}}{\pi_i}]$, $p_{i,i-1} = \frac{1}{2} \min[1, \frac{\pi_{i-1}}{\pi_i}]$, and $p_{i,i} = 1 - p_{i,i+1} - p_{i,i-1}$, with $p_{ij} = 0$ otherwise (where $\pi_j = 0$ if $j \notin S$). An equivalent algorithmic version is: Given X_{n-1}, let Y_n equal $X_{n-1} \pm 1$ (probability $1/2$ each), and let $U_n \sim \text{Uniform}[0, 1]$ as in (A.3.8), with the $\{U_n\}$ chosen i.i.d., and then let

$$
X_n = \begin{cases} Y_n, & U_n \le \frac{\pi_{Y_n}}{\pi_{X_{n-1}}} & (\text{``accept"}) \\ X_{n-1}, & \text{otherwise} & (\text{``reject"}) \end{cases}
$$

which gives the same transition probabilities by (A.3.9).
An animated illustration is at: www.probability.ca/met
The key point of this algorithm is:

(2.6.1) MCMC Convergence Theorem. The above Metropolis algorithm Markov chain has the property that $\lim_{n \to \infty} p_{ij}^{(n)} = \pi_j$, and $\lim_{n \to \infty} \mathbf{P}[X_n = j] = \pi_j$, for all $i, j \in S$ and all initial distributions ν.

Proof: Here $\pi_i p_{i,i+1} = \pi_i \frac{1}{2} \min[1, \frac{\pi_{i+1}}{\pi_i}] = \frac{1}{2} \min[\pi_i, \pi_{i+1}]$.
Also $\pi_{i+1} p_{i+1,i} = \pi_{i+1} \frac{1}{2} \min[1, \frac{\pi_i}{\pi_{i+1}}] = \frac{1}{2} \min[\pi_{i+1}, \pi_i]$.
So $\pi_i p_{ij} = \pi_j p_{ji}$ if $j = i + 1$, and similarly if $j = i - 1$.
Also, $\pi_i p_{ij} = \pi_j p_{ji}$ if $|j - i| \ge 2$, since then both sides are 0.
And, if $i = j$, then always $\pi_i p_{ij} = \pi_j p_{ji}$.
Hence, $\pi_i p_{ij} = \pi_j p_{ji}$ for all $i, j \in S$.
So, the chain is <u>reversible</u> w.r.t. $\{\pi_i\}$. So, $\{\pi_i\}$ is <u>stationary</u>.
Also, the chain is easily checked to be irreducible and aperiodic.
Hence, the result follows from Theorem (2.4.1). ∎

---

[2]First introduced by N. Metropolis, A. Rosenbluth, M. Rosenbluth, A. Teller, and E. Teller, "Equations of state calculations by fast computing machines", J. Chem. Phys. **21**, 1087–1091, way back in 1953 (!). For an overview of research in this area, see e.g. S. Brooks, A. Gelman, G.L. Jones, and X.-L. Meng, eds., "Handbook of Markov chain Monte Carlo", Chapman & Hall / CRC Press, 2011.

Theorem (2.6.1) says that for "large enough" n, X_n is approximately a *sample* from π. (Though again, it doesn't say how large.) Such Markov chains are very widely used to sample from complicated distributions $\{\pi_i\}$, to estimate probabilities and expectations; indeed, "markov chain monte carlo" gives over a million hits in Google! Some of the applications use more general versions of the Metropolis algorithm (see Problem (2.6.4)), or related MCMC algorithms such as the *Gibbs sampler* (see Problem (2.6.6)).
And, some of the applications are on <u>continuous</u> state spaces, where π is a *density function* (e.g. a *Bayesian posterior density*); we shall consider that case in Section 4.8 below.

(2.6.2) Problem. Let $S = \{1, 2, 3\}$, with $\pi_1 = 1/2$ and $\pi_2 = 1/3$ and $\pi_3 = 1/6$. Find (with proof) irreducible transition probabilities $\{p_{ij}\}_{i,j \in S}$ such that π is a stationary distribution.   [sol]

(2.6.3) Problem. Let $S = \mathbf{N}$, with $\pi_i = 2/3^i$ for each $i \in S$. Find (with proof) irreducible transition probabilities $\{p_{ij}\}_{i,j \in S}$ such that π is a stationary distribution.

(2.6.4) Problem. [*General Metropolis algorithm.*]   Let S be a contiguous subset of $\mathbf{Z}$, let π be a probability distribution on S with $\pi_i > 0$ for all $i \in S$, and let $q : S \times S \to \mathbf{R}^+$ be a *proposal distribution* with $\sum_j q(i, j) = 1$ for all $i \in S$. Assume q is *symmetric*, i.e. $q(i, j) = q(j, i)$ for all $i, j \in S$. For $i \neq j$, let $p_{ij} = q(i, j) \min[1, \frac{\pi_j}{\pi_i}]$, and let $p_{ii} = 1 - \sum_{j \neq i} p_{ij}$ be the leftover probabilities. (As before, we take $\pi_j = 0$ if $j \notin S$.)
(a) Prove that (p_{ij}) is a valid Markov chain transition matrix.
(b) Show this Markov chain is reversible with respect to π.
(c) Specify conditions which guarantee that $\lim_{n \to \infty} \mathbf{P}(X_n = j) = \pi_j$ for all $j \in S$. [Hint: Don't forget Theorem (2.4.1).]
(d) Relate your result to Problem (2.4.15).

(2.6.5) Problem. [*Metropolis-Hastings algorithm.*]   Let S, π, and q be as in Problem (2.6.4), except do <u>not</u> assume that q is symmetric, merely that it is symmetrically positive, i.e. $q(i, j) > 0$ <u>iff</u> $q(j, i) > 0$. For $i \neq j$, let $p_{ij} = q(i, j) \min[1, \frac{\pi_j q(j,i)}{\pi_i q(i,j)}]$, and $p_{ii} = 1 - \sum_{j \neq i} p_{ij}$.
(a) Prove that (p_{ij}) is a valid Markov chain transition matrix.

(b) Show this Markov chain is reversible with respect to π.

(c) Specify conditions which guarantee that $\lim_{n\to\infty} \mathbf{P}(X_n = j) = \pi_j$ for all $j \in S$.

(2.6.6) Problem. [*Gibbs sampler.*]   Let $S = \mathbf{Z} \times \mathbf{Z}$, and let $f : S \to (0, \infty)$ be some function from S to the positive real numbers. Let $K = \sum_{(x,y) \in S} f(x, y)$, and assume that $K < \infty$. For $x, y \in \mathbf{Z}$, let $C(x) = \sum_{w \in \mathbf{Z}} f(x, w)$, and $R(y) = \sum_{z \in \mathbf{Z}} f(z, y)$. Consider the following algorithm. Given a pair $(X_{n-1}, Y_{n-1}) \in S$ at each time $n - 1$, it chooses either the "horizontal" or "vertical" option, with probability $1/2$ each. If it chooses horizontal, then it sets $Y_n = Y_{n-1}$, and chooses X_n randomly to equal x with probability $f(x, Y_{n-1}) / R(Y_{n-1})$ for each $x \in \mathbf{Z}$. If it chooses vertical, then it sets $X_n = X_{n-1}$, and chooses Y_n randomly to equal y with probability $f(X_{n-1}, y) / C(X_{n-1})$ for each $y \in \mathbf{Z}$.

(a) Verify that the resulting sequence $\{(X_n, Y_n)\}_{n \in \mathbf{N}}$ has transition probabilities given by

$$
P_{(x,y),(z,w)} = \begin{cases}
\frac{f(z,w)}{2\,C(x)} + \frac{f(z,w)}{2\,R(y)}, & x = z \text{ and } y = w \\[2mm]
\frac{f(z,w)}{2\,C(x)}, & x = z \text{ and } y \neq w \\[2mm]
\frac{f(z,w)}{2\,R(y)}, & x \neq z \text{ and } y = w \\[2mm]
0, & \text{otherwise}
\end{cases}
$$

(b) Verify directly that $\sum_{(z,w) \in S} P_{(x,y),(z,w)} = 1$ for all $(x, y) \in S$.

(c) Show that the chain is reversible with respect to $\pi_{(x,y)} = \frac{f(x,y)}{K}$.

(d) Compute $\lim_{n\to\infty} p^{(n)}_{(x,y),(z,w)}$ for all $x, y, z, w \in \mathbf{Z}$.

2.7. Application – Random Walks on Graphs

Let V be a non-empty finite or countable set.
Let $w : V \times V \to [0, \infty)$ be a weight function which is symmetric (i.e. $w(u, v) = w(v, u)$ for all $u, v \in V$).
The usual *unweighted graph* case is: $w(u, v) = 1$ if there is an edge between u and v, otherwise $w(u, v) = 0$.

But we can also have other weights, multiple edges, self-edges, etc.

Let $d(u) = \sum_{v \in V} w(u, v)$ be the *degree* of the vertex u.
<u>Assume</u> that $d(u) > 0$ for all $u \in V$ (for example, by giving any isolated point a self-edge).
Then we can define a Markov chain as follows:

(2.7.1) Definition. Given a vertex set V and a symmetric weight function w, the *(simple) random walk on the (undirected) graph* (V, w) is the Markov chain with state space $S = V$ and transition probabilities $p_{uv} = \frac{w(u,v)}{d(u)}$ for all $u, v \in V$.

It follows (check!) that $\sum_{v \in V} p_{uv} = \frac{\sum_{v \in V} w(u,v)}{\sum_{v \in V} w(u,v)} = 1$, as it must.

In the usual case where each $w(u, v) = 0$ or 1, from u the chain moves to one of the $d(u)$ vertices connected to u, each with probability $1/d(u)$.

(2.7.2) Ring Graph. Suppose $V = \{1, 2, 3, 4, 5\}$, with $w(i, i+1) = w(i + 1, i) = 1$ for $i = 1, 2, 3, 4$, and $w(5, 1) = w(1, 5) = 1$, with $w(i, j) = 0$ otherwise (see Figure 11).
Is random walk on this graph irreducible? Is it aperiodic? Does it have a stationary distribution? Do its transitions converge?

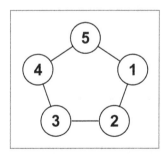

Figure 11: A diagram of the Ring Graph.

(2.7.3) Stick Graph. Suppose $V = \{1, 2, \ldots, K\}$, with $w(i, i+1) = w(i + 1, i) = 1$ for $1 \le i \le K - 1$, with $w(i, j) = 0$ otherwise (see Figure 12).
Does it have a stationary distribution? etc.

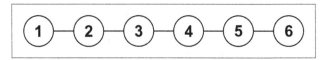

Figure 12: A diagram of the Stick Graph, with $K = 6$.

(2.7.4) Star Graph. Suppose $V = \{0, 1, 2, \ldots, K\}$, with $w(i, 0) = w(0, i) = 1$ for $i = 1, 2, 3$, with $w(i, j) = 0$ otherwise (see Figure 13). Does it have a stationary distribution? Will it converge?

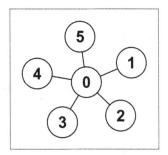

Figure 13: A diagram of the Star Graph, with $K = 5$.

(2.7.5) Infinite Graph. Suppose $V = \mathbf{Z}$, with $w(i, i+1) = w(i+1, i) = 1$ for all $i \in V$, and $w(i, j) = 0$ otherwise (see Figure 14). Random walk on this graph corresponds exactly to simple symmetric random walk.

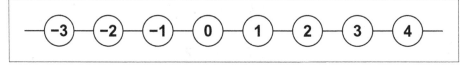

Figure 14: A diagram of part of the Infinite Graph (2.7.5).

(2.7.6) Frog Graph. Suppose $V = \{1, 2, \ldots, K\}$, with $w(i, i) = 1$ for $1 \le i \le K$, and $w(i, i+1) = w(i+1, i) = 1$ for $1 \le i \le K-1$, and $w(K, 1) = w(1, K) = 1$, and $w(i, j) = 0$ otherwise (see Figure 15). When $K = 20$, random walk on this graph corresponds exactly to the original Frog Walk from Section 1.1.

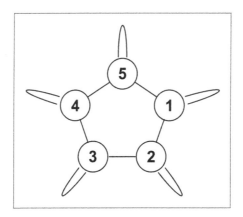

Figure 15: A diagram of the Frog Graph with $K = 5$.

Do these walks have stationary distributions? Usually yes!

Let $Z = \sum_{u \in V} d(u) = \sum_{u,v \in V} w(u,v)$.

In the unweighted case, Z equals two times the number of edges. (Except that self-edges are only counted once, not twice.)

Here Z might be <u>infinite</u>, if V is infinite.

But if $Z < \infty$, then we have a precise formula for a stationary distribution:

(2.7.7) Graph Stationary Distribution. Consider a random walk on an (undirected) graph V with degrees $d(u)$. Assume that Z is <u>finite</u>. Then if $\pi_u = \frac{d(u)}{Z}$ for all $u \in V$, then π is a stationary distribution.

Proof: It is easily checked that $\pi_u \geq 0$, and $\sum_u \pi_u = 1$, so π is a probability distribution on V.

Also, $\pi_u p_{uv} = \frac{d(u)}{Z} \frac{w(u,v)}{d(u)} = \frac{w(u,v)}{Z}$.

And, $\pi_v p_{vu} = \frac{d(v)}{Z} \frac{w(v,u)}{d(v)} = \frac{w(v,u)}{Z} = \frac{w(u,v)}{Z}$. Equal!

So, the chain is reversible with respect to π.

So, by Proposition (2.2.2), π is a stationary distribution. ∎

Thus, (2.7.7) gives a precise formula for the stationary distribution of almost <u>any</u> (simple) random walk on any (undirected) graph.

This is remarkable, since no such formula exists for most other Markov chains.

(However, in e.g. the Infinite Graph (2.7.5) corresponding to s.s.r.w., there is no stationary distribution, and (2.7.7) does not apply, because $Z = \infty$.)

What about irreducibility?

Well, if the graph is *connected* (i.e. it is possible to move from any vertex to any other vertex through edges of positive weight), then the chain is clearly irreducible.

What about periodicity?

Well, the walk can always return to any vertex u in 2 steps, by moving to any other vertex and then back along the same edge.

So, 1 and 2 are the only possible periods.

If the graph is a *bipartite graph* (i.e., it can be divided into two subsets such that all the links go from one subset to the other one), then the chain has period 2.

But if the chain is not bipartite, then it is aperiodic.

Furthermore, if there is any cycle of odd length, then the graph cannot be bipartite, so it must be aperiodic.

We conclude:

(2.7.8) Graph Convergence Theorem. For a random walk on a connected non-bipartite graph, if $Z < \infty$, then $\lim_{n \to \infty} p_{uv}^{(n)} = \frac{d(v)}{Z}$ for all $u, v \in V$, and $\lim_{n \to \infty} \mathbf{P}[X_n = v] = \frac{d(v)}{Z}$ (for any initial probabilities).

Proof: By the above, if it is connected then it is irreducible, and if it is non-bipartite then it is aperiodic.

Also, if $Z < \infty$ then $\pi_v = \frac{d(v)}{Z}$ is a stationary distribution by (2.7.7).

Hence, by the Markov Chain Convergence Theorem (2.4.1), for all $u, v \in V$, $\lim_{n \to \infty} p_{uv}^{(n)} = \pi_v = \frac{d(v)}{Z}$. ∎

What about graphs which might not be bipartite?

Since the only possible periods are 1 and 2, the Periodic Convergence Theorem (2.5.1) and Average Probability Convergence (2.5.2) give:

(2.7.9) Graph Average Convergence.   A random walk on a connected graph with $Z < \infty$ (whether bipartite or not) has $\lim_{n\to\infty} \frac{1}{2}[p_{uv}^{(n)} + p_{uv}^{(n+1)}] = \frac{d(v)}{Z}$, and $\lim_{n\to\infty} \frac{1}{n} \sum_{\ell=1}^{n} p_{uv}^{(\ell)} = \frac{d(v)}{Z}$ for all $u, v \in V$.

Consider again the Stick Graph (2.7.3), with $V = \{1, 2, \ldots, K\}$, and $w(i, i+1) = w(i+1, i) = 1$ for $1 \leq i \leq K-1$, and $w(i, j) = 0$ otherwise.

This graph is clearly connected, but is bipartite (since every edge goes between and even vertex and an odd vertex).

Also $d(u) = 1$ for $i = 1$ or K, otherwise $d(u) = 2$.

Hence, $Z = 1 + 2 + 2 + \ldots + 2 + 1 = 1 + 2(K-2) + 1 = 2K - 2$.

Then if $\pi_i = \frac{d(i)}{Z} = \frac{1}{2K-2}$ for $i = 1$ or K, and $\pi_i = \frac{2}{2K-2}$ for $2 \leq i \leq K-1$, then π is a stationary distribution by (2.7.7).

Then, we must have $\lim_{n\to\infty} \frac{1}{2}[p_{ij}^{(n)} + p_{ij}^{(n+1)}] = \pi_j$ for all $j \in V$.

(2.7.10) Problem. Consider again the Star Graph (2.7.4).
(a) Is this random walk irreducible?
(b) What periods do the states of this random walk have?
(c) Does $\lim_{n\to\infty} p_{uv}^{(n)}$ exist, and if so then what does it equal?
(d) Does $\lim_{n\to\infty} \frac{1}{2}[p_{uv}^{(n)} + p_{uv}^{(n+1)}]$ exist, and if so what does it equal?

(2.7.11) Problem. Repeat Problem (2.7.10) except with an extra edge between 0 and itself, i.e. with $w(0, 0) = 1$ instead of $w(0, 0) = 0$.

(2.7.12) Problem. Consider the undirected graph with vertex set $V = \{1, 2, 3, 4\}$, and an undirected edge (of weight 1) between each of the following four pairs of edges (and no other edges): (1,2), (2,3), (3,4), and (2,4). Let $\{p_{uv}\}_{u,v\in V}$ be the transition probabilities for random walk on this graph. Compute (with full explanation) $\lim_{n\to\infty} p_{21}^{(n)}$, or prove that this limit does not exist.   [sol]

(2.7.13) Problem.   Consider the undirected graph on the vertices $V = \{1, 2, 3, 4, 5\}$, with weights given by $w(1, 2) = w(2, 1) = w(2, 3) = w(3, 2) = w(1, 3) = w(3, 1) = w(3, 4) = w(4, 3) = w(3, 5) = w(5, 3) = 1$, and $w(u, v) = 0$ otherwise.
(a) Draw a picture of this graph.
(b) Compute (with full explanation) $\lim_{n\to\infty} \mathbf{P}[X_n = 3]$, where $\{X_n\}$ is the usual (simple) random walk on this graph.

2.8. Mean Recurrence Times

Another important quantity is the following:

(2.8.1) Definition. The *mean recurrence time* of a state i is $m_i = \mathbf{E}_i(\inf\{n \geq 1 : X_n = i\}) = \mathbf{E}_i(T_i)$ where $T_i = \inf\{n \geq 1 : X_n = i\}$.

That is, m_i is the expected time to return from i back to i.

Now, if the chain <u>never</u> returns to i, then $T_i = \infty$.
If this has positive probability (which holds <u>iff</u> i is transient), then $m_i = \infty$, cf. (A.5.1).
So, if i is transient, then $m_i = \infty$.
It follows that if $m_i < \infty$, then i must be recurrent.
However, even if i is recurrent, i.e. $\mathbf{P}_i(T_i < \infty) = 1$, it is still possible that $\mathbf{E}_i(T_i) = \infty$; cf. Section A.5.
This leads to the following definitions:

(2.8.2) Definition. A state is *positive recurrent* if $m_i < \infty$.
Or, it is *null recurrent* if it is recurrent but $m_i = \infty$.

Is there any connection to stationary distributions? Yes!
Suppose a chain starts at i, and $m_i < \infty$.
Over the long run, what <u>fraction</u> of time will it spend at i?
Well, on average, the chain returns to i once every m_i steps.
That is, for large r, by the Law of Large Numbers (A.7.2), it takes about rm_i steps for the chain to return a total of r times.
So, in a large number $n \approx rm_i$ of steps, it will return to i about r times, i.e. about n/m_i times, i.e. about $1/m_i$ of the time.
Hence, the limiting <u>fraction</u> of time it spends at i is $1/m_i$:

$$(2.8.3) \qquad \lim_{n\to\infty} \frac{1}{n} \sum_{k=1}^{n} 1_{X_k=i} = 1/m_i \quad \text{w.p. 1}.$$

Then by the Bounded Convergence Theorem (A.10.1),

$$(2.8.4) \qquad \lim_{n\to\infty} \mathbf{E}_i\Big(\frac{1}{n} \sum_{k=1}^{n} 1_{X_k=i}\Big) = 1/m_i.$$

However, if the chain is irreducible, with stationary distribution π, then by finite linearity and Average Probability Convergence (2.5.2),

$$\lim_{n\to\infty} \mathbf{E}_i\left(\frac{1}{n}\sum_{k=1}^{n} \mathbf{1}_{X_k=i}\right) = \lim_{n\to\infty} \frac{1}{n}\sum_{k=1}^{n} \mathbf{E}_i(\mathbf{1}_{X_k=i})$$

$$= \lim_{n\to\infty} \frac{1}{n}\sum_{k=1}^{n} \mathbf{P}_i(X_k = i) = \lim_{n\to\infty} \frac{1}{n}\sum_{k=1}^{n} p_{ii}^{(k)} = \pi_i.$$

So, surprisingly, we must have $1/m_i = \pi_i$, i.e. $m_i = 1/\pi_i$.
(This explains the names "positive recurrent" and "null recurrent": they come from considering $1/m_i$, not m_i.)
It follows that $\sum_{i\in S}(1/m_i) = 1$, which seems surprising too.

If $m_i = \infty$, then it follows as above that we must have $\pi_i = 1/m_i = 1/\infty = 0$.
Hence, if $m_j = \infty$ for <u>all</u> j, then we must have $\pi_j = 0$ for all j, which contradicts that $\sum_{j\in S} \pi_j = 1$.
So, in that case, there is <u>no</u> stationary distribution.

In turns out that, similar to (2.2.9), if a chain is irreducible, and $m_j < \infty$ for <u>some</u> $j \in S$, then $m_i < \infty$ for <u>all</u> $i \in S$, and $\{1/m_i\}$ is stationary; see e.g. Section 8.4 of Rosenthal (2006).

Putting this all together, we summarise the results as:

(2.8.5) Recurrence Time Theorem. For an irreducible Markov chain, either (α) $m_i < \infty$ for all $i \in S$, and there is a unique stationary distribution given by $\pi_i = 1/m_i$; or (β) $m_i = \infty$ for all $i \in S$, and there is <u>no</u> stationary distribution.

Furthermore, we have the following fact similar to the Finite Space Theorem (1.6.8) (for a proof see e.g. Rosenthal, 2006, Proposition 8.4.10):

(2.8.6) Proposition. An irreducible Markov chain on a <u>finite</u> state space S always falls into case (α) above, i.e. has $m_i < \infty$ for all $i \in S$, and a unique stationary distribution given by $\pi_i = 1/m_i$.

The converse to Proposition (2.8.6) is false.

For example, Example (2.2.12) has infinite state space $S = \mathbf{N}$, but still has a stationary distribution, so it still falls into case (α).

(2.8.7) Problem. Let $S = \{1, 2\}$ and $P = \begin{pmatrix} 2/3 & 1/3 \\ 3/4 & 1/4 \end{pmatrix}$.

(a) Draw a diagram of this Markov chain. [sol]

(b) Compute m_1 and m_2 directly. [Hint: Remember the geometric distribution (A.3.4).] [sol]

(c) Use m_1 and m_2 to specify a stationary distribution π. [sol]

(d) Verify directly that π is indeed stationary. [sol]

What about simple random walk (s.r.w.)? We have:

(2.8.8) Proposition. Simple random walk, for any $0 < p < 1$, has infinite mean recurrence times, i.e. has $m_i = \infty$ for all $i \in S$.

Proof: We know that s.r.w. is irreducible.

However, s.r.w. has no stationary distribution by (2.2.6).

So, it cannot be in case (α) of Theorem (2.8.5).

Hence, it must be in case (β).

So, we must have $m_i = \infty$ for all i.   ■

That is, on average s.r.w. (including s.s.r.w. when $p = 1/2$) takes an infinite amount of time to return to where it started.

This is surprising, since s.s.r.w. is recurrent, and usually returns quickly. But the small chance of taking a very long time to return is sufficient to make the expected return time infinite even though the actual return time is always finite; cf. Section A.5.

In summary, simple symmetric random walk (s.s.r.w.) is: recurrent and hence in case (a) of the Cases Theorem (1.6.7) with $f_{ij} = 1$ and $\sum_{n=1}^{\infty} p_{ij}^{(n)} = \infty$ for all $i, j \in S$; but in case (i) of the Vanishing Together Corollary (2.2.9) with $\lim_{n \to \infty} p_{ij}^{(n)} = 0$ for all $i, j \in S$ and no stationary distribution; and in case (β) of the Recurrence Time Theorem (2.8.5) with expected return times $m_i = \infty$ for all $i \in S$.

Finally, for s.r.w. we consider $\mathbf{E}_i(T_{i+1})$, i.e. the expected time starting from i to first hit the state $i + 1$.

Well, at time 0, s.r.w. moves from i to either $i+1$ with probability p, or $i-1$ with probability $1-p$, which takes one step. Hence,

$$m_i := \mathbf{E}_i(T_i) = 1 + p\,\mathbf{E}_{i+1}(T_i) + (1-p)\,\mathbf{E}_{i-1}(T_i).$$

Then, by shift-invariance, this is the same as

$$m_i = 1 + p\,\mathbf{E}_i(T_{i-1}) + (1-p)\,\mathbf{E}_i(T_{i+1}).$$

But we know that $m_i = \infty$, so

$$\infty = 1 + p\,\mathbf{E}_i(T_{i-1}) + (1-p)\,\mathbf{E}_i(T_{i+1}).$$

Hence, at least <u>one</u> of $\mathbf{E}_i(T_{i-1})$ and $\mathbf{E}_i(T_{i+1})$ must be infinite. Now, if $p = 1/2$, then by symmetry $\mathbf{E}_i(T_{i+1}) = \mathbf{E}_i(T_{i-1})$, so

$$(\textbf{2.8.9}) \qquad \mathbf{E}_i(T_{i+1}) = \mathbf{E}_i(T_{i-1}) = \infty,$$

i.e. on average it takes an <u>infinite</u> amount of time for s.s.r.w. to reach the state just to the right (or, equivalently, the state just to the left). By contrast, for simple random walk with $p > 1/2$, $\mathbf{E}_i(T_{i+1}) < \infty$, but in that case $f_{i,i-1} < 1$ by (1.7.5), so $\mathbf{P}_i(T_{i-1} = \infty) > 0$, so $\mathbf{E}_i(T_{i-1}) = \infty$ by (A.5.1).

2.9. Application – Sequence Waiting Times

Suppose we repeatedly flip a fair coin, and get Heads (H) or Tails (T) independently each time with probability $1/2$ each.
Let τ be the first time the sequence "HTH" is completed.
What is $\mathbf{E}(\tau)$? And, is the answer the same for "THH"?

Surprisingly, this problem can be solved using Markov chains! Let X_n be the amount of the desired sequence (HTH) that the chain has "achieved" at the nth flip.
(For example, if the flips begin with HHTTHTH, then $X_1 = 1$, $X_2 = 1$, $X_3 = 2$, $X_4 = 0$, $X_5 = 1$, $X_6 = 2$, and $X_7 = 3$, and $\tau = 7$.)
We always have $X_\tau = 3$, since we "win" upon reaching state 3.
Assume we "start over" right after we win, i.e. that $X_{\tau+1} = 1$ if flip $\tau+1$ is Heads, otherwise $X_{\tau+1} = 0$, just like for X_1.

We take $X_0 = 0$, i.e. at the beginning we have not achieved anything. This is equivalent to $X_0 = 3$, since after 3 we start over anyway. But this means that the mean waiting time of "HTH" is equal to the mean recurrence time of the state 3!

Here $\{X_n\}$ is a Markov chain with state space is $S = \{0, 1, 2, 3\}$. What are the transition probabilities?

Well, $p_{01} = p_{12} = p_{23} = 1/2$ (the probability of continuing the sequence).

Also $p_{00} = p_{20} = 1/2$ (the probability of a Tail ruining the sequence). But is $p_{10} = 1/2$? No!

If the sequence is ruined on the second flip with a Head, then we've already taken the first step towards a fresh sequence beginning with H. (That is, if you fail to match the second flip, T, then you've already matched the first flip, H, for the next try. This is key!) Hence, $p_{11} = 1/2$, while $p_{10} = 0$.

Also, since we "start over" right after we win, therefore $p_{3j} = p_{0j}$ for all j, i.e. $p_{31} = p_{30} = 1/2$. Thus, the transitions are

$$P = \begin{pmatrix} 1/2 & 1/2 & 0 & 0 \\ 0 & 1/2 & 1/2 & 0 \\ 1/2 & 0 & 0 & 1/2 \\ 1/2 & 1/2 & 0 & 0 \end{pmatrix}.$$

Using the equation $\pi P = \pi$, it can then be computed (check!) that the stationary distribution is $(0.3, 0.4, 0.2, 0.1)$.

So, by the Recurrence Time Theorem (2.8.5), the mean time to return from state 3 to state 3 is $1/\pi_3 = 1/0.1 = 10$.

But returning from state 3 to state 3 has the same probabilities as going from state 0 to state 3, i.e. as achieving $X_n = 3$ from $X_0 = 0$. Hence, the mean time to go from state 0 to state 3 is also 10.

That is, the mean waiting time for HTH is 10. We've solved it!

What about THH? Is it the same?

Here we compute similarly but differently (check!) that

$$P = \begin{pmatrix} 1/2 & 1/2 & 0 & 0 \\ 0 & 1/2 & 1/2 & 0 \\ 0 & 1/2 & 0 & 1/2 \\ 1/2 & 1/2 & 0 & 0 \end{pmatrix}.$$

Then (check!) the stationary distribution is $(1/8, 1/2, 1/4, 1/8)$.
So, the expected waiting time for "THH" is equal to the mean recurrence time of state 3, which is $1/\pi_3 = 1/(1/8) = 8$.
This is <u>smaller</u> than the previous answer 10.
So, surprisingly[3], HTH and THH are not the same!
(Intuitively, the reason is that the sequence THH is often ruined by getting a Tail, but that first Tail already starts us towards completing the sequence the next time. By contrast, HTH is less often ruined with a Head to begin its next attempt.)

(2.9.1) Problem. Suppose we repeatedly flip a fair coin. Let τ be the number of flips until we first see two Heads in a row, i.e. see "HH". Compute the expected value $\mathbf{E}(\tau)$. [sol]

(2.9.2) Problem. Repeat the previous problem where the coin is now <u>biased</u> (not fair), and instead shows Heads with probability 2/3 or Tails with probability 1/3 (independently on each flip). [Answer: 15/4.]

(2.9.3) Problem. Suppose we repeatedly flip a fair coin. Let τ be the number of flips until we first see three Heads in a row, i.e. see the pattern "HHH". Compute $\mathbf{E}(\tau)$.

(2.9.4) Problem. Suppose we repeatedly flip a coin which is <u>biased</u> (not fair), and shows H with probability 3/4 or T with probability 1/4 (independently on each flip). Let τ be the number of flips until we first see three Heads in a row, i.e. see the pattern "HHH". Compute $\mathbf{E}(\tau)$.

(2.9.5) Problem. Suppose we repeatedly flip a fair coin.
(a) Compute the expected value of the number of flips until we first see the pattern "HTHT".
(b) Repeat part (a) for the pattern "HTTH".

(2.9.6) Problem. Suppose we repeatedly roll a fair six-sided die (which is equally likely to show 1, 2, 3, 4, 5, or 6). Let τ be the

---

[3]If you're skeptical, you could try running a simulation, e.g. in R with the program at: www.probability.ca/Rseqwait

number of rolls until the pattern "565" first appears. Compute $\mathbf{E}[\tau]$. [Answer: 222.]

(2.9.7) Problem. Suppose we repeatedly roll a fair six-sided die. Let τ be the number of rolls until the pattern "556" first appears. Compute $\mathbf{E}[\tau]$. [Answer: 216.]

3. Martingales

Martingales[4] are a model for stochastic processes which "stay the same on average".

For motivation, consider again the Gambler's Ruin problem of Section 1.7, in the case where $p = 1/2$.
Let $T = \inf\{n \geq 0 : X_n = 0 \text{ or } c\}$ be the time the game ends.
We know from (1.7.1) that $\mathbf{P}(X_T = c) = s(a) = a/c$, and $\mathbf{P}(X_T = 0) = 1 - s(a) = 1 - (a/c)$.
We can then compute that $\mathbf{E}(X_T) = c\,\mathbf{P}(X_T = c) + 0\,\mathbf{P}(X_T = 0) = c\,s(a) + 0\,(1 - s(a)) = c\,(a/c) + 0\,(1 - (a/c)) = a$.
So, $\mathbf{E}(X_T) = \mathbf{E}(X_0)$, i.e. "on average it stays the same".
This makes sense since if $p = 1/2$ then $\mathbf{E}(X_{n+1} \,|\, X_n = i) = (1/2)(i + 1) + (1/2)(i - 1) = i$, i.e. each step stays the same on average.

We can also apply the reverse logic:
Suppose we *knew* that the process stayed the same on average, so that $\mathbf{E}(X_T) = \mathbf{E}(X_0) = a$, but we did not know $s(a)$.
We still must have $\mathbf{E}(X_T) = c\,s(a) + 0\,(1 - s(a))$.
It follows that $c\,s(a) + 0\,(1 - s(a)) = \mathbf{E}(X_0) = a$.
Re-arranging, we would solve that $s(a) = a/c$.
This provides a much easier solution to the Gambler's Ruin problem (at least when $p = 1/2$).

Reasoning like this is very powerful.
But can we really be sure that $\mathbf{E}(X_T) = \mathbf{E}(X_0)$, even though T is a random time which depends on the process? We'll see!

3.1. Martingale Definitions

Let $\{X_n\}_{n=0}^{\infty}$ be a sequence of random variables.
We assume throughout that the random variables X_n have *finite expectation* (or, are *integrable*), i.e. that $\mathbf{E}|X_n| < \infty$ for all n. (This allows us to take conditional expectations below.)

---

[4]The name "martingale" apparently comes from the betting strategies discussed in Remark (1.7.7); see e.g. "The origins of the word martingale" by R. Mansuy, Electronic Journal for History of Probability and Statistics **5(1)**, June 2009.

(3.1.1) Definition. A sequence $\{X_n\}_{n=0}^{\infty}$ is a *martingale* if

$$\mathbf{E}(X_{n+1} \mid X_0, ..., X_n) \;=\; X_n, \qquad n = 0, 1, 2, \ldots .$$

Definition (3.1.1) says[5] that no matter what has happened so far, the conditional average of the next value will be equal to the most recent value.

For <u>discrete</u> random variables, the definition means that

(3.1.2) $\mathbf{E}[X_{n+1} \mid X_0 = i_0, ..., X_n = i_n] \;=\; i_n$

for all $i_0, i_1, \ldots, i_n$, in terms of discrete conditional expectation (A.6.3).

If the sequence $\{X_n\}$ is a *Markov chain*, then

$$\mathbf{E}[X_{n+1} \mid X_0 = i_0, ..., X_n = i_n]$$

$$= \sum_{j \in S} j\, P[X_{n+1} = j | X_0 = i_0, ..., X_n = i_n]$$

$$= \sum_{j} j\, P[X_{n+1} = j | X_n = i_n] \;=\; \sum_{j} j\, p_{i_n, j} \,.$$

To be a martingale, this must equal i_n.
That is, a (discrete) Markov chain is a martingale if

(3.1.3) $\displaystyle\sum_{j \in S} j\, p_{ij} \;=\; i \quad \text{for all } i \in S .$

(3.1.4) Problem. Let $S = \{1, 2, 3, 4\}$, with $p_{11} = p_{44} = 1$, and $p_{21} = p_{23} = 1/2$, and $p_{32} = p_{33} = 0$. Find values of p_{31} and p_{34} which make this Markov chain be a martingale. **[sol]**

Let $\{X_n\}$ be s.s.r.w. (i.e., simple random walk with $p = 1/2$), with state space $S = \mathbf{Z}$. Is it a martingale?

---

[5]Technical Remark: Martingales are often defined instead by the condition that $\mathbf{E}(X_{n+1} \mid \mathcal{F}_n) = X_n$ for some nested "filtration" $\{\mathcal{F}_n\}$, i.e. sub-σ-algebras $\{\mathcal{F}_n\}$ with $\sigma(X_0, X_1, \ldots, X_n) \subseteq \mathcal{F}_n \subseteq \mathcal{F}_{n+1}$. But it follows from a version of the double-expectation formula that this condition is actually equivalent to Definition (3.1.1); see e.g. Remark 14.0.1 of Rosenthal (2006). I thank Neal Madras for a very helpful discussion of these issues.

Well, we always have $|X_n| \leq n$, so $\mathbf{E}|X_n| \leq n < \infty$, so there is no problem[6] regarding finite expectations. (Indeed, we will almost always have $\mathbf{E}|X_n| < \infty$, so we usually won't even mention it.) More importantly, for all $i \in S$, we compute that $\sum_{j \in S} j\, p_{ij} = (i+1)(1/2) + (i-1)(1/2) = i$.
This satisfies (3.1.3), so s.s.r.w. is indeed a martingale. (In fact, it is my favourite one!)

If $\{X_n\}$ is a martingale, then by the *Double-Expectation Formula* (A.6.6),

$$\mathbf{E}(X_{n+1}) \;=\; \mathbf{E}\Big[\mathbf{E}(X_{n+1} \mid X_0, X_1, \ldots, X_n)\Big] \;=\; \mathbf{E}(X_n),$$

i.e.

(3.1.5) $\mathbf{E}(X_n) \;=\; \mathbf{E}(X_0)$ for all n.

This is not surprising, since martingales stay the same on average. However, (3.1.5) is not a sufficient condition for $\{X_n\}$ to be a martingale, as the following problem shows:

(3.1.6) Problem. Let $\{X_n\}$ be a Markov chain with state space $S = \mathbf{Z}$, and $X_0 = 0$, and $p_{0,1} = p_{0,-1} = 1/2$, and $p_{i,i+1} = 1$ for all $i \geq 1$, and $p_{i,i-1} = 1$ for all $i \leq -1$.
(a) Draw a diagram of this Markov chain.
(b) Compute $\mathbf{P}(X_n = i)$ for each $i \in S$.
(c) Use this to compute $\mathbf{E}(X_n)$ for all $n \in \mathbf{N}$.
(d) Verify that (3.1.5) is satisfied.
(e) Show that (3.1.3) is not satisfied in this case, e.g. when $i = 1$.
(f) Compute $\mathbf{P}(X_2 = i \mid X_0 = 0, X_1 = 1)$ for all $i \in S$.
(g) Use this to compute $\mathbf{E}(X_2 \mid X_0 = 0, X_1 = 1)$.
(h) Show that (3.1.2) is not satisfied in this case, i.e. that $\{X_n\}$ is not a martingale.

(3.1.7) Problem. (*) Consider a Markov chain $\{X_n\}$ with state space $S = \mathbf{Z}$, with $X_0 = 0$, and with transition probabilities given

---

[6]Aside: For s.s.r.w., the $\mathbf{E}|X_n|$ are not uniformly bounded, i.e. there is no constant $M < \infty$ such that $\mathbf{E}|X_n| < M$ for all n at once. But this is not required for a martingale. All we require is that $\mathbf{E}|X_n| < \infty$ for each fixed n, a much weaker condition.

by $p_{00} = 0$, $p_{0j} = c/|j|^3$ for all $j \neq 0$, and $p_{i,i+i^2} = p_{i,i-i^2} = 1/2$
for all $i \neq 0$, with $p_{ij} = 0$ otherwise, where $c > 0$ is chosen so that
$\sum_{j \in S} p_{0j} = 1$. For each of the following statements, determine (with
proof) whether it is true or false. [Hint: Don't forget (A.4.2) that
$\sum_{k=1}^{\infty} (1/k^a)$ is finite iff $a > 1$.]
(a) $\sum_{j \in S} j\, p_{ij} = i$ for all $i \in S$.
(b) $\mathbf{E}|X_1| < \infty$.
(c) $\mathbf{E}|X_2| < \infty$.
(d) $\{X_n\}$ is a martingale.

3.2. Stopping Times and Optional Stopping

For martingales, we know from (3.1.5) that $\mathbf{E}(X_n) = \mathbf{E}(X_0)$ for
each fixed time n.

However, we will often want to consider $\mathbf{E}(X_T)$ for a <u>random</u> time T
(which might depend on the process itself).

Will we still have $\mathbf{E}(X_T) = \mathbf{E}(X_0)$, even if T is random?

One issue is, we need to prevent the random time T from looking
into the <u>future</u> of the process, before deciding whether to stop.

This prompts the following definition.

(3.2.1) Definition. A non-negative-integer-valued random variable
T is a *stopping time* for $\{X_n\}$ if the event $\{T = n\}$ is determined
by $X_0, X_1, \ldots, X_n$, i.e. if the indicator function $\mathbf{1}_{T=n}$ is a function of
$X_0, X_1, \ldots, X_n$.

Intuitively, Definition (3.2.1) says that a stopping time T must
decide whether to stop at time n based solely on what has happened
up to time n, i.e. without first looking into the future.

Some examples can help to clarify this:

e.g. $T = 5$ <u>is</u> a valid stopping time.

e.g. $T = \inf\{n \geq 0 : X_n = 5\}$ <u>is</u> a valid stopping time. (Note: here
$T = \infty$ if $\{X_n\}$ never hits 5.)

e.g. $T = \inf\{n \geq 0 : X_n = 0 \text{ or } X_n = c\}$ <u>is</u> a valid stopping time.

e.g. $T = \inf\{n \geq 2 : X_{n-2} = 5\}$ <u>is</u> a valid stopping time.

e.g. $T = \inf\{n \geq 2 : X_{n-1} = 5, X_n = 6\}$ <u>is</u> a valid stopping time.

e.g. $T = \inf\{n \geq 0 : X_{n+1} = 5\}$ is <u>not</u> a valid stopping time, since it looks into the future to stop one step <u>before</u> reaching 5.

Another issue is, what if $\mathbf{P}(T = \infty) > 0$?
Then X_T isn't always <u>defined</u>, so we can't say if $\mathbf{E}(X_T) = \mathbf{E}(X_0)$.
To avoid that, we assume T is a stopping time with $P(T < \infty) = 1$.
<u>Then</u> do we always have $\mathbf{E}(X_T) = \mathbf{E}(X_0)$? Not necessarily!

For example, let $\{X_n\}$ be s.s.r.w. with $X_0 = 0$.
We already know that this is a martingale.
Now, let $T = T_{-5} = \inf\{n \geq 0 : X_n = -5\}$.
Is T a stopping time? Yes indeed.
And, $\mathbf{P}(T < \infty) = 1$ since s.s.r.w. is recurrent.
However, in this case, we always have $X_T = -5$.
So, $\mathbf{E}(X_T) = -5 \neq 0 = \mathbf{E}(X_0)$. Why not? What went wrong?
It turns out that we need some <u>additional</u> conditions.
One possibility is <u>boundedness</u>, leading to our first of several Optional Stopping (sometimes called *Optional Sampling*) results:

(3.2.2) Optional Stopping Lemma. If $\{X_n\}$ is a martingale, and T is a stopping time which is <u>bounded</u> (i.e., $\exists M < \infty$ with $\mathbf{P}(T \leq M) = 1$), then $\mathbf{E}(X_T) = \mathbf{E}(X_0)$.

Proof: Using a telescoping sum, and then indicator functions, and then finite linearity of expectation (A.9.1),

$$\mathbf{E}(X_T) - \mathbf{E}(X_0) = \mathbf{E}(X_T - X_0) = \mathbf{E}\left[\sum_{k=1}^{T}(X_k - X_{k-1})\right]$$

$$= \mathbf{E}\left[\sum_{k=1}^{M}(X_k - X_{k-1})\mathbf{1}_{k \leq T}\right] = \sum_{k=1}^{M}\mathbf{E}\left[(X_k - X_{k-1})\mathbf{1}_{k \leq T}\right]$$

$$= \sum_{k=1}^{M}\mathbf{E}\left[(X_k - X_{k-1})(1 - \mathbf{1}_{T \leq k-1})\right].$$

Next, using the *Double-Expectation Formula* (A.6.6), and then the fact that "$1 - \mathbf{1}_{T \leq k-1}$" is completely determined by $X_0, X_1, \ldots, X_{k-1}$

and thus can be conditionally factored out as in (A.6.7), it follows that

$$\mathbf{E}(X_T) - \mathbf{E}(X_0)$$

$$= \sum_{k=1}^{M} \mathbf{E}\left(\mathbf{E}\left[(X_k - X_{k-1})(1 - \mathbf{1}_{T \le k-1}) \;\middle|\; X_0, X_1, \ldots, X_{k-1}\right]\right)$$

$$= \sum_{k=1}^{M} \mathbf{E}\left((1 - \mathbf{1}_{T \le k-1}) \, \mathbf{E}\left[(X_k - X_{k-1}) \;\middle|\; X_0, X_1, \ldots, X_{k-1}\right]\right)$$

$$= \sum_{k=1}^{M} \mathbf{E}\left((1 - \mathbf{1}_{T \le k-1}) \, \mathbf{E}\left[X_k \;\middle|\; X_0, X_1, \ldots, X_{k-1}\right]\right)$$

$$- \sum_{k=1}^{M} \mathbf{E}\left((1 - \mathbf{1}_{T \le k-1}) \, \mathbf{E}\left[X_{k-1} \;\middle|\; X_0, X_1, \ldots, X_{k-1}\right]\right).$$

But $\mathbf{E}\left[X_k \;\middle|\; X_0, X_1, \ldots, X_{k-1}\right] = X_{k-1}$ since $\{X_n\}$ is a martingale, and $\mathbf{E}\left[X_{k-1} \;\middle|\; X_0, X_1, \ldots, X_{k-1}\right] = X_{k-1}$ by (A.6.8) since X_{k-1} is a function of $X_0, X_1, \ldots, X_{k-1}$. Hence, $\mathbf{E}(X_T) - \mathbf{E}(X_0)$ equals

$$\sum_{k=1}^{M} \mathbf{E}\left((1 - \mathbf{1}_{T \le k-1})(X_{k-1})\right) - \sum_{k=1}^{M} \mathbf{E}\left((1 - \mathbf{1}_{T \le k-1})(X_{k-1})\right) = 0,$$

so $\mathbf{E}(X_n) = \mathbf{E}(X_0)$, as claimed. ∎

(3.2.3) **Problem.** Explain why the above proof fails if $M = \infty$.

(3.2.4) **Example.** Consider s.s.r.w. with $X_0 = 0$, and let

$$T = \min\left(10^{12}, \; \inf\{n \ge 0 : X_n = -5\}\right).$$

Then T is a stopping time. Also, $T \le 10^{12}$, so T is bounded. By the Optional Stopping Lemma (3.2.2), $\mathbf{E}(X_T) = \mathbf{E}(X_0) = 0$. But nearly always (i.e., whenever the process hits -5 some time within the first 10^{12} steps), we will have $X_T = -5$.

How can this be? How can a random variable like X_T which is nearly always equal to -5, still have expected value equal to 0?
Well, by the *Law of Total Expectation* (A.6.4),

$$0 \ = \ \mathbf{E}(X_T) \ = \ \mathbf{P}(X_T = -5)\, \mathbf{E}(X_T \,|\, X_T = -5)$$

$$+ \, \mathbf{P}(X_T \neq -5)\, \mathbf{E}(X_T \,|\, X_T \neq -5)\,.$$

Here $\mathbf{E}(X_T \,|\, X_T = -5) = -5$.
Also, here $q := \mathbf{P}(X_T = -5) \approx 1$, so $\mathbf{P}(X_T \neq 5) = 1 - q \approx 0$.
So, the equation becomes

$$0 \ = \ q(-5) + (1 - q)\mathbf{E}(X_T \,|\, X_T \neq -5)$$

where $q \approx 1$. How is this possible?
We must have that $\mathbf{E}(X_T \,|\, X_T \neq -5) = 5q/(1 - q)$ is <u>huge</u>.
This is plausible since, if the process did <u>not</u> hit -5 within the first 10^{12} steps, then probably it instead got very, very large.

Can we apply the Optional Stopping Lemma to the Gambler's Ruin problem? No, since there T is not bounded (though still finite, cf. Section A.5). Instead, we need a more general result:

(3.2.5) Optional Stopping Theorem. If $\{X_n\}$ is a martingale with stopping time T, and $\mathbf{P}(T < \infty) = 1$, and $\mathbf{E}|X_T| < \infty$, and if $\lim_{n \to \infty} \mathbf{E}(X_n \mathbf{1}_{T>n}) = 0$, then $\mathbf{E}(X_T) = \mathbf{E}(X_0)$.

Proof: For each $m \in \mathbf{N}$, let $S_m = \min(T, m)$.
Then S_m is a stopping time, and it is bounded (by m).
So, by the Optional Stopping Lemma, $\mathbf{E}(X_{S_m}) = \mathbf{E}(X_0)$.
But $X_{S_m} = X_{\min(T,m)} = X_T \mathbf{1}_{T \leq m} + X_m \mathbf{1}_{T>m}$,
i.e. $X_{S_m} = X_T(1 - \mathbf{1}_{T>m}) + X_m \mathbf{1}_{T>m} = X_T - X_T \mathbf{1}_{T>m} + X_m \mathbf{1}_{T>m}$.
So, $X_T = X_{S_m} + X_T \mathbf{1}_{T>m} - X_m \mathbf{1}_{T>m}$.
Hence, $\mathbf{E}(X_T) = \mathbf{E}(X_{S_m}) + \mathbf{E}(X_T \mathbf{1}_{T>m}) - \mathbf{E}(X_m \mathbf{1}_{T>m})$.
So, by the above, $\mathbf{E}(X_T) = \mathbf{E}(X_0) + \mathbf{E}(X_T \mathbf{1}_{T>m}) - \mathbf{E}(X_m \mathbf{1}_{T>m})$.
This is true for <u>any</u> m. Then take $m \to \infty$.
We have that $\lim_{m \to \infty} \mathbf{E}(X_m \mathbf{1}_{T>m}) = 0$ by assumption.
   Also, $\lim_{m \to \infty} \mathbf{E}(X_T \mathbf{1}_{T>m}) = 0$ *by the Dominated Convergence Theorem (A.10.3), since* $\mathbf{E}|X_T| < \infty$ *and* $\mathbf{1}_{T>m} \to 0$.

So, $\mathbf{E}(X_T) \to \mathbf{E}(X_0)$, i.e. $\mathbf{E}(X_T) = \mathbf{E}(X_0)$, as claimed.   ∎

Can we apply <u>this</u> result to the Gambler's Ruin problem? Almost – we just need one more corollary first.

(3.2.6) Optional Stopping Corollary. If $\{X_n\}$ is a martingale with stopping time T, which is "bounded up to time T" (i.e., $\exists M < \infty$ with $\mathbf{P}(|X_n|\mathbf{1}_{n \le T} \le M) = 1 \ \forall n$), and $\mathbf{P}(T < \infty) = 1$, then $\mathbf{E}(X_T) = \mathbf{E}(X_0)$.

Proof: First, it clearly follows that $\mathbf{P}(|X_T| \le M) = 1$.
   [Formally, $\mathbf{P}(|X_T| > M) = \sum_n \mathbf{P}(T = n, |X_T| > M) = \sum_n \mathbf{P}(T = n, |X_n|\mathbf{1}_{n \le T} > M) \le \sum_n \mathbf{P}(|X_n|\mathbf{1}_{n \le T} > M) = \sum_n (0) = 0.]$
Hence, $\mathbf{E}|X_T| \le M < \infty$. Also,

$$|\mathbf{E}(X_n \mathbf{1}_{T>n})| \ \le \ \mathbf{E}(|X_n|\mathbf{1}_{T>n}) \ = \ \mathbf{E}(|X_n|\mathbf{1}_{n \le T}\mathbf{1}_{T>n})$$

$$\le \ \mathbf{E}(M\mathbf{1}_{T>n}) = M\,\mathbf{P}(T > n),$$

which $\to 0$ as $n \to \infty$ since $\mathbf{P}(T < \infty) = 1$.
The result then follows from the Optional Stopping Theorem.   ∎

Consider again Gambler's Ruin from Section 1.7, with $p = 1/2$. Let $T = \inf\{n \ge 0 : X_n = 0 \text{ or } X_n = c\}$ be the time the game ends. Then $\mathbf{P}(T < \infty) = 1$ by Proposition (1.7.6).
Also, if the game has <u>not</u> yet ended, i.e. $n \le T$, then X_n must be between 0 and c. Hence, $|X_n|\mathbf{1}_{n \le T} \le c < \infty$ for all n.
So, by the Optional Stopping Corollary, $\mathbf{E}(X_T) = \mathbf{E}(X_0) = a$.
Hence, $a = c\,s(a) + 0\,(1 - s(a))$, so we must have $s(a) = a/c$.
(This is a much easier solution than in Section 1.7.)

What about Gambler's Ruin with $p \ne 1/2$?
Then is $\{X_n\}$ a martingale?
<u>No</u>, since $\sum_j j\,p_{ij} = p(i+1) + (1-p)(i-1) = i + 2p - 1 \ne i$.
Instead, we use a trick: Let $Y_n = \left(\frac{1-p}{p}\right)^{X_n}$.
Then $\{Y_n\}$ is also a Markov chain. And,

$$\mathbf{E}(Y_{n+1} \mid Y_0, Y_1, \ldots, Y_n)$$

$$= p\left(\frac{1-p}{p}\right)^{X_n+1} + (1-p)\left(\frac{1-p}{p}\right)^{X_n-1}$$

$$= p\left[Y_n\left(\frac{1-p}{p}\right)\right] + (1-p)\left[Y_n \Big/ \left(\frac{1-p}{p}\right)\right]$$

$$= Y_n(1-p) + Y_n(p) = Y_n.$$

So, $\{Y_n\}$ is a martingale!

And, again $\mathbf{P}(T < \infty) = 1$ by Proposition (1.7.6).

And, $|Y_n|1_{n\leq T} \leq \max\left[\left(\frac{1-p}{p}\right)^0, \left(\frac{1-p}{p}\right)^c\right] < \infty$ for all n.

Hence, by the Optional Stopping Corollary, $\mathbf{E}(Y_T) = \mathbf{E}(Y_0) = \left(\frac{1-p}{p}\right)^a$.

But $Y_T = \left(\frac{1-p}{p}\right)^c$ if we win, or $Y_T = \left(\frac{1-p}{p}\right)^0 = 1$ if we lose.

Hence, $\left(\frac{1-p}{p}\right)^a = s(a)\left(\frac{1-p}{p}\right)^c + [1-s(a)](1) = 1 + s(a)\left[\left(\frac{1-p}{p}\right)^c - 1\right]$.

Solving, $s(a) = \frac{\left(\frac{1-p}{p}\right)^a - 1}{\left(\frac{1-p}{p}\right)^c - 1}$, as before. (Again, a much easier solution.)

In summary, martingales provide a much easier solution to the Gambler's Ruin problem, whether $p = 1/2$ or $p \neq 1/2$.

(3.2.7) Problem. Let $\{X_n\}$ be a Markov chain on the state space $S = \{1,2,3,4\}$, with $X_0 = 3$, and with transition probabilities $p_{11} = p_{44} = 1$, $p_{21} = 1/3$, $p_{34} = 1/4$, and $p_{23} = p_{31} = p_{12} = p_{13} = p_{14} = p_{41} = p_{42} = p_{43} = 0$. Let $T = \inf\{n \geq 0 : X_n = 1 \text{ or } 4\}$.
(a) Find non-negative values of p_{22}, p_{24}, p_{32}, and p_{33} so $\sum_{j\in S} p_{ij} = 1$ for all $i \in S$, and also $\{X_n\}$ is a martingale. [sol]
(b) For the values found in part (a), compute $\mathbf{E}(X_T)$. [sol]
(c) For the values in part (a), compute $\mathbf{P}(X_T = 1)$. [sol]

(3.2.8) Problem. Let $\{X_n\}$ be a Markov chain on the state space $S = \{1,2,3,4\}$, with $X_0 = 2$, and transition probabilities $p_{11} = p_{44} = 1$, $p_{21} = 1/4$, $p_{34} = 1/5$, $p_{23} = p_{31} = p_{12} = p_{13} = p_{14} = p_{41} = p_{42} = p_{43} = 0$. Let $T = \inf\{n \geq 0 : X_n = 1 \text{ or } 4\}$.
(a) Find non-negative values of p_{22}, p_{24}, p_{32}, and p_{33} so $\sum_{j\in S} p_{ij} = 1$ for all $i \in S$, and also $\{X_n\}$ is a martingale. [sol]
(b) For the values found in part (a), compute $\mathbf{E}(X_T)$. [sol]
(c) For the values in part (a), compute $\mathbf{P}(X_T = 1)$. [sol]

(3.2.9) Problem. Let $\{X_n\}$ be a Markov chain on the state space $S = \{1, 2, 3, \ldots\}$ of all positive integers, which is also a martingale. Assume $X_0 = 5$, and that there is $c > 0$ such that $p_{i,i-1} = c$ and $p_{i,i+2} = 1 - c$ for all $i \geq 2$. Let $T = \inf\{n \geq 0 : X_n = 1 \text{ or } X_n \geq 10\}$.
(a) Determine (with explanation) what c must equal. [Hint: Remember that $\{X_n\}$ is a martingale.] [sol]
(b) Determine (with explanation) what p_{11} must equal. [Hint: Again, remember that $\{X_n\}$ is a martingale.] [sol]
(c) Determine (with explanation) the value of $\mathbf{E}(X_3)$. [sol]
(d) Determine (with explanation) the value of $\mathbf{E}(X_T)$. [sol]
(e) Prove or disprove that $\sum_{n=1}^{\infty} p_{55}^{(n)} = \infty$. [sol]

(3.2.10) Problem. Let $\{Z_i\}_{i=1}^{\infty}$ be an i.i.d. collection of random variables with $\mathbf{P}[Z_i = -1] = 3/4$ and $\mathbf{P}[Z_i = C] = 1/4$, for some $C > 0$. Let $X_0 = 5$, and $X_n = 5 + Z_1 + Z_2 + \ldots + Z_n$ for $n \geq 1$. Finally, let $T = \inf\{n \geq 1 : X_n = 0 \text{ or } Z_n > 0\}$.
(a) Find a value of C such that $\{X_n\}$ is a martingale.
(b) For this value of C, compute (with explanation) $\mathbf{E}(X_9)$.
(c) For this value of C, compute (with explanation) $\mathbf{E}(X_T)$. [Hint: Is T bounded?]

(3.2.11) Problem. Consider simple symmetric random walk $\{X_n\}$ with $X_0 = 0$. For $m \in \mathbf{Z}$, let $T_m = \inf\{n \geq 1 ; X_n = m\}$, and let $U = \min(T_4, T_{-6})$. Prove or disprove each of the following statements:
(a) $\mathbf{E}[X_U] = 0$.
(b) $\mathbf{E}[X_{T_4}] = 0$.

3.3. Wald's Theorem

We begin with a warm-up example:

(3.3.1) Example. Suppose we repeatedly roll a fair six-sided die (which is equally likely to show 1, 2, 3, 4, 5, or 6).
Let Z_n be the result of the nth roll.
Let $R = \inf\{n \geq 1 : Z_n = 5\}$ be the first time we roll 5.
Let A be the sum of all the numbers rolled up to time R, i.e. $A = \sum_{n=1}^{R} Z_n$.
What is $\mathbf{E}(A)$?

Also, let $S = \inf\{n \geq 1 : Z_n = 3\}$ be the first time we roll 3, and $B = \sum_{n=1}^{S} Z_n$ be the sum up to time S.

Which is larger, $\mathbf{E}(A)$ or $\mathbf{E}(B)$? Are they equal?

Finally, let $A' = \sum_{n=1}^{R-1} Z_n$ and $B' = \sum_{n=1}^{S-1} Z_n$ be the sums not counting the final roll.

Then which is larger, $\mathbf{E}(A')$ or $\mathbf{E}(B')$? Are they equal?

To answer questions like this, we shall use:

(3.3.2) Wald's Theorem. Suppose $X_n = a + Z_1 + \ldots + Z_n$, where $\{Z_i\}$ are i.i.d., with finite mean m. Let T be a stopping time for $\{X_n\}$ which has finite mean, i.e. $\mathbf{E}(T) < \infty$. Then $\mathbf{E}(X_T) = a + m\,\mathbf{E}(T)$.

Proof: We compute that

$$\mathbf{E}(X_T) - a = \mathbf{E}(X_T - a) = \mathbf{E}(Z_1 + \ldots + Z_T)$$

$$= \mathbf{E}\left[\sum_{i=1}^{T} Z_i\right] = \mathbf{E}\left[\sum_{i=1}^{\infty} Z_i\,\mathbf{1}_{T \geq i}\right] = \mathbf{E}\left[\lim_{N \to \infty} \sum_{i=1}^{N} Z_i\,\mathbf{1}_{T \geq i}\right].$$

Assuming we can interchange the expectation and limit, then

$$\mathbf{E}(X_T) - a = \lim_{N \to \infty} \mathbf{E}\left[\sum_{i=1}^{N} Z_i\,\mathbf{1}_{T \geq i}\right]$$

$$= \lim_{N \to \infty} \sum_{i=1}^{N} \mathbf{E}\left[Z_i\,\mathbf{1}_{T \geq i}\right] = \sum_{i=1}^{\infty} \mathbf{E}\left[Z_i\,\mathbf{1}_{T \geq i}\right].$$

Now, $\{T \geq i\} = \{T \leq i - 1\}^C$, so since T is a stopping time, Z_i is independent of the event $\{T \geq i\}$. Hence,

$$\mathbf{E}(X_T) - a = \sum_{i=1}^{\infty} \mathbf{E}[Z_i]\,\mathbf{E}[\mathbf{1}_{T \geq i}] = \sum_{i=1}^{\infty} m\,\mathbf{P}[T \geq i]$$

$$= m\sum_{i=1}^{\infty} \mathbf{P}[T \geq i] = m\sum_{i=1}^{\infty} \mathbf{E}[T],$$

where the last equality uses the trick (A.2.1).

It remains to justify interchanging the above expectation and limit. This follows from the Dominated Convergence Theorem (A.10.3)

with sequence $X_N = \sum_{i=1}^{N} Z_i \, \mathbf{1}_{T \geq i}$ *and limit* $X = \sum_{i=1}^{\infty} Z_i \, \mathbf{1}_{T \geq i}$
and dominator $Y = \sum_{i=1}^{\infty} |Z_i| \, \mathbf{1}_{T \geq i}$ *since then* $|X_N| \leq Y$ *for all* N
(by the triangle inequality), while by (A.9.2) and (A.2.1), $\mathbf{E}(Y) =$
$\mathbf{E}[\sum_{i=1}^{\infty} |Z_i| \, \mathbf{1}_{T \geq i}] = \sum_{i=1}^{\infty} \mathbf{E}[|Z_i| \, \mathbf{1}_{T \geq i}] = \sum_{i=1}^{\infty} \mathbf{E}|Z_i| \, \mathbf{E}[\mathbf{1}_{T \geq i}] =$
$\mathbf{E}|Z_1| \sum_{i=1}^{\infty} \mathbf{P}[T \geq i] = \mathbf{E}|Z_1| \, \mathbf{E}(T) < \infty.$ ∎

A special case is when $m = 0$. Then, $\{X_n\}$ is a martingale, and
Wald's Theorem says that $\mathbf{E}(X_T) = a = \mathbf{E}(X_0)$, as usual, but under
different assumptions (an i.i.d. sum, with $\mathbf{E}(T) < \infty$).

Consider the example where $\{X_n\}$ is s.s.r.w. with $X_0 = 0$, and
$T = \inf\{n \geq 0 : X_n = -5\}$ is the first time we hit -5.
Then in Wald's Theorem, $a = 0$ and $m = 0$.
Also, $\mathbf{P}(T < \infty) = 1$ since s.s.r.w. is recurrent.
But here $X_T = -5$, so $\mathbf{E}(X_T) = -5 \neq 0 = \mathbf{E}(X_0)$.
Does this contradict Wald's Theorem?
No, since here $\mathbf{E}(T) = \infty$, as shown in (2.8.9).

Let's apply Wald's Theorem to the dice Example (3.3.1).
In this case, the $\{Z_i\}$ are i.i.d. with mean $m = 3.5$.
And, clearly R and S are stopping times.
Also, R and S have Geometric$(1/6)$ distributions as in (A.3.4), so
$\mathbf{E}(R) = \mathbf{E}(S) = 6$.
In the notation of Wald's Theorem, $a = 0$, and $A = X_R$, and $B = X_S$.
So, $\mathbf{E}(A) = \mathbf{E}(X_R) = a + m\,\mathbf{E}(R) = 3.5(6) = 21$.
Similarly, $\mathbf{E}(B) = \mathbf{E}(X_S) = a + m\,\mathbf{E}(S) = 3.5(6) = 21$. Equal!

What about A' and B'?
Well, $A' = X_{R-1}$ and $B' = X_{S-1}$. However, $R-1$ and $S-1$ are not
stopping times, so we cannot apply Wald's Theorem.
On the other hand, clearly $A' = A - 5$, and $B' = B - 3$.
So, $\mathbf{E}(A') = \mathbf{E}(A) - 5 = 21 - 5 = 16$.
And, $\mathbf{E}(B') = \mathbf{E}(B) - 3 = 21 - 3 = 18$. So, B' is larger!

Next, we apply Wald's Theorem to the Gambler's Ruin problem
of Section 1.7, to compute the average time the game takes.
Assume first that $p \neq 1/2$, and let $T = \inf\{n \geq 0 : X_n = 0 \text{ or } c\}$.
Then what is $\mathbf{E}(T) =$ expected number of bets in the game?
We have the following:

(3.3.3) Proposition. If $\{X_n\}$ is Gambler's Ruin with $p \neq 1/2$, and $T = \inf\{n \geq 0 : X_n = 0 \text{ or } c\}$, then

$$\mathbf{E}(T) = \frac{1}{2p-1}\left(c\frac{\left(\frac{1-p}{p}\right)^a - 1}{\left(\frac{1-p}{p}\right)^c - 1} - a\right).$$

Proof: Here $Z_i = +1$ if you win the ith bet, otherwise $Z_i = -1$.
So, $m = \mathbf{E}(Z_i) = p(1) + (1-p)(-1) = 2p - 1 \neq 0$.
And, $\mathbf{E}(T) < \infty$ by Proposition (1.7.6).
Hence, by Wald's Theorem, $\mathbf{E}(X_T) = a + m\,\mathbf{E}(T)$, so
$\mathbf{E}(T) = \frac{1}{m}\left(\mathbf{E}(X_T) - a\right)$. But here

$$\mathbf{E}(X_T) = c\,s(a) + 0\,(1 - s(a)) = c\frac{\left(\frac{1-p}{p}\right)^a - 1}{\left(\frac{1-p}{p}\right)^c - 1}.$$

So, $\mathbf{E}(T) = \frac{1}{m}\left(\mathbf{E}(X_T) - a\right) = \frac{1}{2p-1}\left(c\frac{\left(\frac{1-p}{p}\right)^a-1}{\left(\frac{1-p}{p}\right)^c-1} - a\right).$ ∎

For example, if $p = 0.49$, $a = 9,700$, and $c = 10,000$, then $\mathbf{E}(T) = 484,997$, i.e. on average the game requires nearly half a million bets!

What about $\mathbf{E}(T)$ when $p = 1/2$?
If $p = 1/2$, then $m = 0$, so the above method does not work.
However, we can still solve for $\mathbf{E}(T)$:

(3.3.4) Proposition. If $\{X_n\}$ is Gambler's Ruin with $p = 1/2$, and $T = \inf\{n \geq 0 : X_n = 0 \text{ or } c\}$, then $\mathbf{E}(T) = \mathbf{Var}(X_T) = a(c - a)$.

To prove Proposition 3.3.4, we need to use a "second order" martingale argument, as follows.

(3.3.5) Proposition. Let $X_n = a + Z_1 + \ldots + Z_n$, where $\{Z_i\}$ are i.i.d. with mean 0 and variance $v < \infty$. Let $Y_n = (X_n - a)^2 - nv = (Z_1 + \ldots + Z_n)^2 - nv$. Then $\{Y_n\}$ is a martingale.

Proof: First, $\mathbf{E}|Y_n| \leq \mathbf{Var}(X_n) + nv = 2nv < \infty$.
Then, since Z_{n+1} is independent of $Z_1, \ldots, Z_n, Y_0, \ldots, Y_n$, with $\mathbf{E}[Z_{n+1}] = 0$ and $\mathbf{E}[Z_{n+1}]^2 = v$, we have

$$\mathbf{E}[Y_{n+1} \,|\, Y_0, Y_1, \ldots, Y_n]$$

$$= \mathbf{E}\Big[(Z_1 + \ldots + Z_n + Z_{n+1})^2 - (n+1)v \ \Big| \ Y_0, Y_1, \ldots, Y_n\Big]$$

$$= \mathbf{E}\Big[(Z_1 + \ldots + Z_n)^2 + (Z_{n+1})^2 + 2\,Z_{n+1}(Z_1 + \ldots + Z_n)$$

$$-nv - v \ \Big| \ Y_0, Y_1, \ldots, Y_n\Big]$$

$$= \mathbf{E}\Big[Y_n + (Z_{n+1})^2 - v + 2\,Z_{n+1}(Z_1 + \ldots + Z_n)$$

$$\Big| \ Y_0, Y_1, \ldots, Y_n\Big]$$

$$= Y_n + v - v + 2\,\mathbf{E}(Z_{n+1})\,\mathbf{E}\Big[Z_1 + \ldots + Z_n \ \Big| \ Y_0, Y_1, \ldots, Y_n\Big]$$

$$= Y_n + v - v + 0 = Y_n . \qquad \blacksquare$$

Proof of Proposition 3.3.4:
Let $Y_n = (X_n - a)^2 - n$ (since here $v = 1$).
Then $\{Y_n\}$ is a martingale by Proposition (3.3.5).
Choose $M > 0$, and let $S_M = \min(T, M)$.
Then S_m is a stopping time, which is bounded by M.
Hence, by the Optional Stopping Lemma (3.2.2), $\mathbf{E}[Y_{S_M}] = \mathbf{E}[Y_0] = (a-a)^2 - 0 = 0$.
But $Y_{S_M} = (X_{S_M} - a)^2 - S_M$, so $\mathbf{E}(S_M) = \mathbf{E}[(X_{S_M} - a)^2]$.
As $M \to \infty$, S_M increases monotonically to T, so $\mathbf{E}(S_M) \to \mathbf{E}(T)$ by the Monotone Convergence Theorem (A.10.2).
Also, $\mathbf{E}[(X_{S_M} - a)^2] \to \mathbf{E}[(X_T - a)^2]$ by the Bounded Convergence Theorem (A.10.1), since $(X_{S_M} - a)^2 \le \max\big(a^2, (c-a)^2\big) < \infty$.
Hence, letting $M \to \infty$ gives $\mathbf{E}(T) = \mathbf{E}[(X_T - a)^2]$.
And, since $\mathbf{E}(X_T) = a$, $\mathbf{E}[(X_T - a)^2] = \mathbf{Var}(X_T)$.
Finally, we compute that $\mathbf{Var}(X_T) = (a/c)\,(c-a)^2 + (1 - a/c)a^2 = (a/c)(c^2 + a^2 - 2ac) + (a^2 - a^3/c) = ac + a^3/c - 2a^2 + a^2 - a^3/c = ac - a^2 = a(c-a)$. $\qquad \blacksquare$

For example, if $c = 10,000$, $a = 9,700$, and $p = 1/2$, then $\mathbf{E}(T) = a(c-a) = 2,910,000$. Even larger!

(3.3.6) Problem. (*) Prove that the formula for $\mathbf{E}(T)$ for Gambler's Ruin is continuous as $p \to 1/2$. [sol]

(3.3.7) Example. Consider again the "double 'til you win" betting strategy discussed in Remark (1.7.7).

Let $p \leq 1/2$ be the probability of winning each bet.
Let $\{Z_i\}$ be i.i.d., with $\mathbf{P}[Z_i = +1] = p$ and $\mathbf{P}[Z_i = -1] = 1 - p$.
Then $m := \mathbf{E}(Z_i) = p - (1 - p) = 2p - 1 \leq 0$.
Suppose we bet 2^{i-1} dollars on the ith bet.
Then our winnings after n bets equals $X_n = \sum_{i=1}^{n} 2^{i-1} Z_i$.
Let $T = \inf\{n \geq 1 : Z_n = +1\}$ be time of our first win, and suppose we $\underline{\text{stop}}$ betting at time T.
Then T has a *geometric distribution* (A.3.4), with $\mathbf{E}(T) = 1/p < \infty$.
Also, w.p. 1, $T < \infty$ and $X_T = -1 - 2 - 4 - \ldots - 2^{T-1} + 2^T = +1$.
That is, we are $\underline{\text{guaranteed}}$ to be up \$1 at time T, even though we start at $a = 0$ and have average gain $m \leq 0$ on each bet.
So, if X_n is the total amount won by time n, then $\mathbf{E}(X_T) = +1$.
However, $a + m \, \mathbf{E}(T) = (2p - 1)/p \leq 0$.
Does this contradict Wald's Theorem? No, since even though $\{Z_i\}$ is i.i.d., the sequence $\{2^{i-1} Z_i\}$ is $\underline{\text{not}}$ i.i.d.
(And, as discussed in Remark (1.7.7), this strategy is $\underline{\text{not}}$ a guaranteed money-maker, due to the possibility of reaching your credit limit and being forced to stop betting with a huge loss.)

3.4. Application – Sequence Waiting Times (Revisited)

Suppose that, as in Section 2.9, we repeatedly flip a fair coin, and let τ be the number of flips until we first see the pattern HTH. Again we can ask, what is the expected time $\mathbf{E}(\tau)$?
We previously solved this problem using mean recurrence times.
We can also solve it using martingales!

Specifically, suppose that at each time n, a new "player" appears, and bets \$1 on Heads, then if they win they bet \$2 on Tails, then if they win again they bet \$4 on Heads.
Each player stops betting as soon as they either lose once (and hence are down a total of \$1), or win three bets in a row (and hence are up a total of \$7). (See Figure 16 for an example.)

Let X_n be the total amount won by $\underline{\text{all}}$ players up to time n.
Then since the bets were fair, $\{X_n\}$ is a $\underline{\text{martingale}}$.
And, τ is a $\underline{\text{stopping time}}$ for $\{X_n\}$.

n	1	2	3	4	5	6	7	8	Total
Flip	H	H	T	T	H	H	T	H	
Player 1	+1	−2	0	0	0	0	0	0	−1
Player 2	0	+1	+2	−4	0	0	0	0	−1
Player 3	0	0	−1	0	0	0	0	0	−1
Player 4	0	0	0	−1	0	0	0	0	−1
Player 5	0	0	0	0	+1	−2	0	0	−1
Player 6	0	0	0	0	0	+1	+2	+4	+7
Player 7	0	0	0	0	0	0	−1	0	−1
Player 8	0	0	0	0	0	0	0	+1	+1
Grand Total									**+2**

Figure 16: Player bet payoffs for HTH when $\tau = 8$.

Here each of the first $\tau - 3$ players (i.e., all but the last three) each has a net loss of \$1 (after either one or two or three bets).
Player $\tau - 2$ wins all three bets, for a gain of \$7.
Player $\tau - 1$ bets on Heads and loses, for a loss of \$1.
Player τ bets on Heads and wins, for a gain of \$1.
Hence, $X_\tau = (\tau - 3)(-1) + (7) + (-1) + (1) = -\tau + 10.$
We next show that $\mathbf{E}(X_\tau) = \mathbf{E}(X_0) = 0$:
> Indeed, if $T_m = \min(\tau, m)$, then $\mathbf{E}(X_{T_m}) = 0$ by the Optional Stopping Lemma (3.2.2), and $\lim_{m \to \infty} \mathbf{E}(X_{T_m}) = \mathbf{E}(X_\tau)$ by the Dominated Convergence Theorem (A.10.3) with dominator $Y = 7\tau$ since $|X_n - X_{n-1}| \le 7$ and $\mathbf{E}(\tau) < \infty.$

Hence, $0 = \mathbf{E}(X_\tau) = \mathbf{E}(-\tau + 10)$, whence $\mathbf{E}(\tau) = 10.$
This is the same answer we got before!
But easier: there was no stationary distribution to compute.

Using a similar approach for THH, we get (check!) that $X_\tau = -(\tau - 3) + (7) + (-1) + (-1) = -\tau + 8$, whence $\mathbf{E}(\tau) = 8$, also the same as before.

(3.4.1) Problem. Use the martingale method to prove that the expected waiting time for a fair coin to show the pattern "TTT" is 14. [sol]

(3.4.2) Problem. Suppose we repeatedly flip a fair coin. Use the martingale method of this section to compute $\mathbf{E}(\tau)$, where τ is the number of flips until we first see the pattern:

(a) HH.
(b) HHH.
(c) HTHT.
(d) HTTH.

3.5. Martingale Convergence Theorem

Suppose $\{X_n\}$ is a martingale.
Then $\{X_n\}$ could have infinite fluctuations in both directions, as we have seen for s.s.r.w. (1.6.14).
Or, $\{X_n\}$ could converge w.p. 1 to some (perhaps random) value, as in the following two examples.

(3.5.1) Example. Let $\{X_n\}$ be Gambler's Ruin with $p = 1/2$, where we <u>stop</u> as soon as we either win or lose. Then $X_n \to X$ w.p. 1, where $\mathbf{P}(X = c) = a/c$ and $\mathbf{P}(X = 0) = 1 - a/c$.

(3.5.2) Example. Let $\{X_n\}$ be a Markov chain on $S = \{2^m : m \in \mathbf{Z}\}$, with $X_0 = 1$, and $p_{i,2i} = 1/3$ and $p_{i,i/2} = 2/3$ for $i \in S$.
This is a martingale, since $\sum_j j\, p_{ij} = (2i)(1/3) + (i/2)(2/3) = i$.
What happens in the long run?
Trick: Let $Y_n = \log_2 X_n$. Then $Y_0 = 0$, and $\{Y_n\}$ is s.r.w. with $p = 1/3$, so $Y_n \to -\infty$ w.p. 1 by the Law of Large Numbers (A.7.2).
Hence, $X_n = 2^{Y_n} \to 2^{-\infty} = 0$ w.p. 1.
That is, $X_n \to X$ w.p. 1, where $\mathbf{P}(X = 0) = 1$.

It turns out that these are the only two possibilities.
That is, if a martingale does <u>not</u> have infinite fluctuations in both directions, i.e. if it is <u>bounded</u> on at least one side, then it must converge w.p. 1:

(3.5.3) Martingale Convergence Theorem. Any martingale $\{X_n\}$ which is bounded below (i.e. $X_n \ge c$ for all n, for some finite number c), <u>or</u> is bounded above (i.e. $X_n \le c$ for all n, for some finite number c), converges w.p. 1 to some random variable X.

The intuition behind this theorem is:

Since the martingale is bounded on one side, it cannot "spread out" forever. And, since it is a martingale, it cannot "drift" forever. So, it has nowhere to go, and eventually has to stop somewhere.

We do not prove this theorem here; for a proof[7] see e.g. Section 14.2 of Rosenthal (2006), or Williams (1991), or Billingsley (1995), or many other advanced probability books.

Consider some examples:
We know that s.s.r.w. is a martingale which does not converge. However, it is not bounded below or above.
If we modify s.s.r.w. to stop whenever it reaches 0 (after starting at a positive number), then it is still a martingale, and now it is non-negative and hence bounded below. But since $f_{i0} = 1$ for all i, it will eventually reach 0, and hence it converges to $X = 0$.
Or, if we instead modify s.s.r.w. so that from 0 it always moves to 1, then it is again non-negative, but it does not converge. However, this modification is not a martingale.
Or, if we instead modify s.s.r.w. to make smaller and smaller (but still symmetric) jumps when it is near 0, to avoid ever going negative, then it is still a martingale, and is non-negative. But then it will converge to some value X (perhaps random, or perhaps 0, depending on how we have modified the jumps).
Or, if we consider s.r.w. with $p = 2/3$ stopped at 0, then it is again non-negative, and if $X_0 = i > 0$ then since $f_{i0} < 1$ it might never reach 0, so it might fail to converge. However, it is not a martingale.
So, if you try to modify a martingale to make it bounded above or below, then the modification will either converge, or will no longer be a martingale.

(3.5.4) Problem. Let $S = \{5, 6, 7, 8, \ldots\}$. Let $\{X_n\}$ be a Markov chain on S with $X_0 = 10$, and transition probabilities $p_{55} = 1$, and $p_{i,i-1} = p_{i,i+1}$ for $i \geq 6$, otherwise $p_{ij} = 0$.
(a) Prove that $\{X_n\}$ is a martingale. [sol]

---

[7]The Martingale Convergence Theorem is usually proven for *non-negative* martingales, i.e. assuming $X_n \geq 0$, i.e. taking $c = 0$. However, our more general version follows from this, since if $X_n \geq c$ then $\{X_n - c\}$ is a non-negative martingale, or if $X_n \leq c$ then $\{-X_n + c\}$ is a non-negative martingale, and in either case the non-negative martingale converges iff $\{X_n\}$ converges.

(b) Determine whether or not $\{X_n\}$ converges w.p. 1 to some random variable X, and if so what X equals.   [sol]

(3.5.5) Problem. Let $S = \{5, 6, 7, 8, \ldots\}$. Let $\{X_n\}$ be a Markov chain on S with $X_0 = 10$, and transition probabilities given by $p_{55} = 1$, and $p_{65} = p_{67} = 1/2$, and for $i \geq 7$, $p_{i,i-1} = p_{i,i-2} = 2/7$ and $p_{i,i+2} = 3/7$, with $p_{i,j} = 0$ otherwise. (You may take as given that $E|X_n| < \infty$ for each n.)
(a) Determine whether or not $\{X_n\}$ is a martingale.
(b) Determine whether or not $\{X_n\}$ converges w.p. 1 to some random variable X, and if so what X equals.

(3.5.6) Problem. Consider a Markov chain $\{X_n\}$ with state space $S = \mathbf{N} \cup \{2^{-m} : m \in \mathbf{N}\}$ (i.e., the set of all positive integers together with all the negative integer powers of 2). Suppose the transition probabilities are given by $p_{2^{-m}, 2^{-m-1}} = 2/3$ and $p_{2^{-m}, 2^{-m+1}} = 1/3$ for all $m \in \mathbf{N}$, and $p_{1, 2^{-1}} = 2/3$ and $p_{1,2} = 1/3$, and $p_{i,i-1} = p_{i,i+1} = 1/2$ for all $i \geq 2$, with $p_{i,j} = 0$ otherwise. Let $X_0 = 2$. [You may assume without proof that $E|X_n| < \infty$ for all n.] And, let $T = \inf\{n \geq 1 : X_n = 2^{-2} \text{ or } 5\}$.
(a) Prove that $\{X_n\}$ is a martingale.
(b) Determine whether or not $E(X_n) = 2$ for each fixed $n \in \mathbf{N}$.
(c) Compute (with explanation) $E(X_T)$.
(d) Compute $P(X_T = 5)$.
(e) Prove $\{X_n\}$ converges w.p. 1 to some random variable X.
(f) For this random variable X, determine $P(X = x)$ for all x.
(g) Determine whether or not $E(X) = \lim_{n \to \infty} E(X_n)$.

(3.5.7) Problem. (*Pòlya's Urn*) A bag contains $r > 0$ Red balls and $b > 0$ Blue balls. Each minute, a ball is drawn from the bag, and then it is returned to the bag together with another ball of the same colour. So, after n minutes, there are a total of $r + b + n$ balls in the bag, of which some random number R_n are Red, and the remaining $B_n = r + b + n - R_n$ are Blue. (Here $R_0 = r$ and $B_0 = b$.) Let $Y_n = R_n/(n + r + b)$ be the fraction of Red balls after n minutes.
(a) Prove that $\{Y_n\}$ is a martingale.
(b) Prove that $Y_n \to Y$ w.p. 1 for some random variable Y.

(3.5.8) Problem. (*) Let $X_1, X_2, \ldots$ be independent random variables with

$$X_n = \begin{cases} 1 & \text{with probability } (2n)^{-1} \\ 0 & \text{with probability } 1 - n^{-1} \\ -1 & \text{with probability } (2n)^{-1}. \end{cases}$$

Let $Y_1 = X_1$, and for $n \geq 2$,

$$Y_n = \begin{cases} X_n & \text{if } Y_{n-1} = 0 \\ nY_{n-1}|X_n| & \text{if } Y_{n-1} \neq 0. \end{cases}$$

(a) Determine whether or not $\{Y_n\}$ is a martingale.
(b) Determine if there is a random variable Y with $Y_n \to Y$ w.p. 1.
(c) Relate the answers in parts (a) and (b) to the Martingale Convergence Theorem.

3.6. Application – Branching Processes

A *branching process* is defined as follows.
Let X_n equal the number of individuals (e.g., people with colds, or bacteria, or ...) which are present at time n.
Start with $X_0 = a$ individuals, for some $0 < a < \infty$.
Then, at each time n, each of the X_n individuals creates a random number of offspring to appear at time $n + 1$ (and then vanishes).
The number of offspring of each individual is i.i.d. $\sim \mu$, where μ is any probability distribution on $\{0, 1, 2, \ldots\}$, called the *offspring distribution*.
That is, each individual has i children with probability $\mu\{i\}$.
(Note: We assume that there is just one "parent" per offspring, i.e. we assume *asexual reproduction*.)
Hence, $X_{n+1} = Z_{n,1} + Z_{n,2} + \ldots + Z_{n,X_n}$, where $\{Z_{n,i}\}_{i=1}^{X_n}$ are i.i.d. $\sim \mu$.
For example, if we start with $X_0 = 2$ individuals, and the first individual has 1 offspring while the second individual has 4 offspring, then at time 1 we would have $X_1 = 1 + 4 = 5$, and so on. (See Figure 17.)
Here $\{X_n\}$ is Markov chain, on the state space $S = \{0, 1, 2, \ldots\}$.

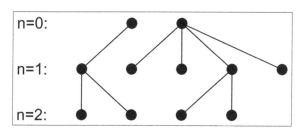

Figure 17: A branching process with $X_0 = 2$, $X_1 = 5$, $X_2 = 4$.

What are the transition probabilities?

Well, $p_{00} = 1$, and $p_{0j} = 0$ for all $j \geq 1$. That is, if X_n ever reaches 0, then it stays there forever (this is called *extinction*).

But p_{ij} for other i is more complicated: p_{ij} is the probability that a sum of i different random variables which are i.i.d. $\sim \mu$ is equal to j. (That is, $p_{ij} = (\mu * \mu * \ldots * \mu)(j)$, a *convolution* of i copies of μ.)

An important question for a branching process is, will it eventually go extinct? That is, will $X_n = 0$ for some n?

Let $m = \sum_i i\,\mu\{i\}$ be the mean of μ, called the *reproductive number*. What relevance does this number have?

Well, assume $0 < m < \infty$. Then

$$\mathbf{E}(X_{n+1} \mid X_0, \ldots, X_n)$$

$$= \mathbf{E}(Z_{n,1} + Z_{n,2} + \ldots + Z_{n,X_n} \mid X_0, \ldots, X_n) = m\,X_n.$$

So, by induction, $\mathbf{E}(X_n) = m^n \mathbf{E}(X_0) = m^n a$.

What does this tell us?

Well, if $m < 1$, then $\mathbf{E}(X_n) = a\,m^n \to 0$. But

$$\mathbf{E}(X_n) = \sum_{k=0}^{\infty} k\,\mathbf{P}(X_n = k) \geq \sum_{k=1}^{\infty} \mathbf{P}(X_n = k) = \mathbf{P}(X_n \geq 1).$$

Hence, $\mathbf{P}(X_n \geq 1) \leq \mathbf{E}(X_n) = a\,m^n \to 0$, i.e. $\mathbf{P}(X_n = 0) \to 1$. That is, if $m < 1$, then extinction is certain!

If $m > 1$, then $\mathbf{E}(X_n) = a\,m^n \to \infty$.

This suggests, and in fact it is true, that in this case, $\mathbf{P}(X_n \to \infty) > 0$, i.e. there is a positive probability that X_n converges to infinity (called *flourishing*).

But assuming $\mu\{0\} > 0$, we still also have $\mathbf{P}(X_n \to 0) > 0$ (for example, this will happen if no one has any offspring at all on the first iteration, which has probability $(\mu\{0\})^a > 0$).

So, $\mathbf{P}(X_n \to \infty) \leq 1 - \mathbf{P}(X_n \to 0) < 1$.

So, if $m > 1$ and $\mu\{0\} > 0$, then we have possible extinction, but also possible flourishing, each with positive probability.

What about the case where $m = 1$?

Then $\mathbf{E}(X_n) = \mathbf{E}(X_0) = a$ for all n.

Here $\mathbf{E}(X_{n+1}|X_0, ..., X_n) = mX_n = X_n$, i.e. $\{X_n\}$ is a martingale.

Also, it is clearly non-negative.

Hence, by the Martingale Convergence Theorem, we must have $X_n \to X$ w.p. 1, for some random variable X.

But how can $\{X_n\}$ converge w.p. 1?

After reaching X, the process $\{X_n\}$ would still continue to fluctuate, i.e. would not converge w.p. 1, unless either (a) $\mu\{1\} = 1$ (i.e. the branching process is *degenerate*), or (b) $X = 0$.

Conclusion: If μ is non-degenerate (i.e., $\mu\{1\} < 1$), then $X \equiv 0$, i.e. $\{X_n\} \to 0$ w.p. 1.

So, in the non-degenerate case, there is certain extinction, even in the borderline case when $m = 1$.

We summarise the above results as:

(3.6.1) Proposition. Consider a branching process, with reproductive number m. Assume that $\mu\{0\} > 0$ (so $\mu\{1\} < 1$ too). Then extinction is certain if $m \leq 1$, but both flourishing and extinction are possible if $m > 1$.

(3.6.2) Problem. Consider a branching process with initial value $X_0 = 1$, and offspring distribution $\mu\{0\} = 4/7$, $\mu\{1\} = 2/7$, and $\mu\{2\} = 1/7$.
(a) Compute $\mathbf{P}(X_1 = 0)$. [sol]
(b) Compute $\mathbf{P}(X_1 = 2)$. [sol]
(c) Compute $\mathbf{P}(X_2 = 2)$. [sol]

(3.6.3) Problem. Consider a branching process with initial value $X_0 = 2$, and with offspring distribution given by $\mu\{0\} = 2/3$ and $\mu\{2\} = 1/3$. Let $T = \inf\{n \geq 1 : X_n = 0\}$ be the extinction time.
(a) Compute $\mathbf{P}(X_1 = 0)$.
(b) Compute $\mathbf{P}(X_1 = 2)$.
(c) Compute $\mathbf{P}(X_2 = 8)$.
(d) Determine whether or not $\mathbf{P}(T < \infty) = 1$.

3.7. Application – Stock Options (Discrete)

In mathematical finance, it is common to model the price of one share of some *stock* as a random process.
For now, we work in discrete time, and suppose that X_n is the price of one share of the stock at each time n.
If you buy the stock, then the situation is clear: if X_n increases then you will make a profit, but if X_n decreases then you will suffer a loss. A more interesting situation arises with stock *options*:

(3.7.1) Definition. A *(European call) stock option* is the <u>option</u> to buy one share of a stock for a fixed *strike price* (or, *exercise price*) $\$K$ at a fixed future *strike time* (or, *maturity time*) $S > 0$.

Such stock options are commonly traded in actual markets; indeed, many millions of dollars are spent on them each day.

Now, if at the strike time S, the stock price X_S is <u>less</u> than the strike price K, the option would <u>not</u> be exercised, and would thus be worth zero.
But if the stock price X_S is <u>more</u> than K, then the option would be exercised to obtain a stock worth X_S for a price of just K, for a net profit of $X_S - K$.
Hence, at time S, the stock option is worth $\max(0, X_S - K)$.

But at time 0 (when the option is to be purchased), X_S is an unknown (random) quantity.
So, we can ask: at time 0, what is the *fair price* of the stock option? Is it the <u>expected value</u> of its worth at time S, i.e. $\mathbf{E}[\max(0, X_S - K)]$? No! It turns out that this is <u>not</u> the fair price.

Instead, the fair price is defined in terms of the concept of *arbitrage*, i.e. a portfolio of investments which guarantees you a positive profit no matter how the stock performs.

Specifically, the fair price of a stock option is defined to be the *no-arbitrage price*, i.e. the price for the option which makes it impossible to make a guaranteed profit through any combination of buying or selling the option, and/or buying and selling the stock.

It is not immediately clear that such a price exists, nor that it is unique. But it turns out that, under appropriate conditions, it is.

In what follows, we <u>assume</u> that you have the ability to buy or sell arbitrary amounts of stock and/or the option at any (discrete) time, without incurring any transaction fees. We begin with an example.

(3.7.2) Example. Suppose at time 0 the stock price X_0 is equal to 100, and at time S the stock price X_S is random with $\mathbf{P}(X_S = 80) = 9/10$ and $\mathbf{P}(X_S = 130) = 1/10$.

Consider an <u>option</u> to buy the stock at time S for $K = 110$.

What is the fair (no-arbitrage) price of this option?

Well, at time S, the option is worth $\max(0, X_S - K)$, which is equal to 0 if $X_S = 80$, or 20 if $X_S = 130$.

Hence, the option's expected value at time S is given by $\mathbf{E}(\text{option}) = \mathbf{E}[\min(0, X_S - K)] = (9/10)(0) + (1/10)(20) = 2$.

Does the option's fair price equal this expected value? Is it more? less? Does it even exist?

Suppose that at time 0, you buy x stock shares for $100 each, and y option shares for some price $c each.

Note: We allow for the possibility that x or y is <u>negative</u>, corresponding to selling (i.e. *shorting*) the stock or option instead of buying positive amounts.

Then if the stock goes up to $130, then you make $130 - $100 = $30 on each stock share, and make $20 - $c = $(20 - c)$ on each option share, for a total profit of $(130 - 100)x + (20 - c)y = 30x + (20 - c)y$.

But if the stock instead goes down to $80, you lose $20 on each stock share, and lose $c on each option share, for a total profit of $(80 - 100)x + (0 - c)y = -20x - cy$.

To attempt to make a <u>guaranteed</u> profit, regardless of how the stock performs, we could attempt to make these two different total profit amounts <u>equal</u> to each other.

Indeed, if $y = -(5/2)x$, then these profits <u>both</u> equal $(5/2)(c-8)x$, i.e. your net profit is no longer random.

So, if $c > 8$, then if you buy $x > 0$ stock shares and $y = -(5/2)x < 0$ option shares, you will make a guaranteed profit $(5/2)(c-8)x > 0$. Or, if $c < 8$, then if you buy $x < 0$ stock shares and $y = -(5/2)x > 0$ option shares, you will make a guaranteed profit $(5/2)(8-c)(-x) > 0$. Either way, there is arbitrage, i.e. a guaranteed profit.

But if $c = 8$, then these both equal 0, so there is <u>no</u> arbitrage. More generally, if $c = 8$, there is no choice of x and y which makes both possible profits positive.

So, for Example (3.7.2), there is no arbitrage <u>iff</u> $c = 8$. This means that $c = \$8$ is the unique fair (no-arbitrage) price. (Not $\mathbf{E}(X_S) = 85$, nor $\mathbf{E}(X_S - K) = -15$, nor even $\mathbf{E}(\text{option}) = \mathbf{E}[\min(0, X_S - K)] = (9/10)(0) + (1/10)(20) = 2$.)

(3.7.3) Problem. Suppose a stock is worth \$50 today, and tomorrow will be worth either \$40 or \$80. Compute the fair (no-arbitrage) price for an option to buy the stock tomorrow for \$70.   [sol]

In Example (3.7.2), what is the connection to martingales? Well, suppose we assign <u>new</u> probabilities $\mathbf{P}(X_S = 80) = 3/5$ and $\mathbf{P}(X_S = 130) = 2/5$, instead of the true probabilities 9/10 and 1/10. (These are called the *Martingale probabilities*.)

Then, for <u>these</u> probabilities, the stock price is a <u>martingale</u> since $(3/5)(80) + (2/5)(130) = 100 = $ initial price, and also the option is a <u>martingale</u> since $(3/5)(0) + (2/5)(130 - 110) = 8 = c = $ initial price. Then, the fair price <u>equals</u> the martingale expected value (i.e. the expected value with respect to the martingale probabilities), i.e. it equals $(3/5)(0) + (2/5)(130 - 110) = 8$.

(In fact, the original probabilities 9/10 and 1/10 are irrelevant, and do not affect the option's value at all!)

This connection is actually true much more generally:

(3.7.4) Martingale Pricing Principle. The fair price of an option equals its expected value under the <u>martingale</u> probabilities, i.e. under the probabilities which make the stock price a martingale.

(3.7.5) Problem. Suppose a stock costs $20 today, and will cost either $10 or $50 tomorrow. Compute the fair (no-arbitrage) price of an option to buy the stock tomorrow for $30, in two ways:
(a) By directly computing the potential profits, as above. [sol]
(b) By using the Martingale Pricing Principle. [sol]

(3.7.6) Problem. Suppose at time 0 the stock price X_0 is equal to 10, and at time S the stock price X_S is random with $\mathbf{P}(X_S = 7) = 2/5$ and $\mathbf{P}(X_S = 11) = 3/5$. Consider the option to buy one share of the stock at time S for price 9.
(a) Find martingale probabilities $q_1 = \mathbf{P}(X_S = 7)$ and $q_2 = \mathbf{P}(X_S = 11)$ which make the stock price a martingale.
(b) Use the Martingale Pricing Principle to compute the fair (no-arbitrage) price of this option.

By similar reasoning, we now solve a more general case:

(3.7.7) Proposition. Suppose a stock price at time 0 equals $X_0 = a$, and at time $S > 0$ equals either $X_S = d$ (down) or $X_S = u$ (up), where $d < a < u$. Then if $d < K < u$, then at time 0, the fair (no-arbitrage) price of an option to buy the stock at time S for K is equal to: $(a - d)(u - K)/(u - d)$.

Proof #1 – Profit Computation: Suppose you buy x shares of the stock, plus y shares of the option.
Then if the stock goes down to $X_S = d$, your profit is $x(d-a)+y(-c)$.
If the stock goes up to $X_S = u$, your profit is $x(u-a)+y(u-K-c)$.
These are equal if $x(d - u) = y(u - K)$, i.e. $y = x(d - u)/(u - K) = -x(u - d)/(u - K)$, in which case your guaranteed profit is $x(d - a) - yc = x(d - a) + x(u - d)/(u - K)c$.
If there is no arbitrage, then this equals zero, which implies that $c = (a - d)(u - K)/(u - d)$.   ∎

Proof #2 – Martingale Pricing Principle: We need to find martingale probabilities $q_1 = \mathbf{P}(X_S = d)$ and $q_2 = \mathbf{P}(X_S = u)$ to make the stock price a martingale.
So, we need $dq_1+uq_2 = a$, i.e. $dq_1+u(1-q_1) = a$, i.e. $(d-u)q_1+u = a$, i.e. $q_1 = (u - a)/(u - d)$.

Then $q_2 = 1 - q_1 = (a - d)/(u - d)$.

Then, by the Martingale Pricing Principle, the fair price is the martingale expectation of the option's worth, which equals $q_1(0) + q_2(u - K) = [(a - d)/(u - d)] * (u - K)$, the same as before.   ∎

(3.7.8) Problem. For the setup of Problem (3.7.5), use Proposition (3.7.7) to verify the answer obtained previously. [sol]

(3.7.9) Problem. For the setup of Example (3.7.2), use Proposition (3.7.7) to verify the answer obtained previously.

Similar (but messier) calculations work in the multi-step discrete case too, and the Martingale Pricing Principle still holds; see e.g. Durrett (2011). And, we shall see later in Section 4.2 that the Martingale Pricing Principle works in <u>continuous</u> time, too.

4. Continuous Processes

So far, we have mostly considered <u>discrete</u> processes, where time is an integer, and the state space is finite or countable.
We now briefly consider a number of different generalisations of this to settings in which time and/or space are <u>continuous</u>.
We begin with a continuous generalisation of simple symmetric random walk (s.s.r.w.), called *Brownian motion*.

4.1. Brownian Motion

Let $\{X_n\}_{n=0}^\infty$ be s.s.r.w., with $X_0 = 0$.
Represent this (as in the proof of (1.6.17)) as $X_n = Z_1 + Z_2 + \ldots + Z_n$, where $\{Z_i\}$ are i.i.d. with $\mathbf{P}(Z_i = +1) = \mathbf{P}(Z_i = -1) = 1/2$.
That is, $X_0 = 0$, and $X_{n+1} = X_n + Z_{n+1}$.
Here $\mathbf{E}(Z_i) = 0$ and $\mathbf{Var}(Z_i) = 1$.

Let M be a large integer, and let $\{Y_t^{(M)}\}$ be like $\{X_n\}$, except with time sped up by a factor of M, and space shrunk down by a factor of $\sqrt{M}$.
That is, $Y_0^{(M)} = 0$, and $Y_{\frac{i+1}{M}}^{(M)} = Y_{\frac{i}{M}}^{(M)} + \frac{1}{\sqrt{M}} Z_{i+1}$, filled in by linear interpolation (see Figure 18).

Intuitively, *Brownian motion* $\{B_t\}_{t \geq 0}$ is the limit as $M \to \infty$ of the processes $\{Y_t^{(M)}\}$, and inherits many of its properties:

First, since $Y_0^{(M)} = 0$ for all M, therefore also $B_0 = 0$.

Second, note[8] that $Y_t^{(M)} = \frac{1}{\sqrt{M}}(Z_1 + Z_2 + \ldots + Z_{tM})$.
It follows that $\mathbf{E}(Y_t^{(M)}) = 0$, and $\mathbf{Var}(Y_t^{(M)}) = (\frac{1}{\sqrt{M}})^2(tM) = t$.
Furthermore, as $M \to \infty$, the Central Limit Theorem (A.7.3) says that $Y_t^{(M)} \to \text{Normal}(0, t)$.
Therefore, also $B_t \sim \text{Normal}(0, t)$.
That is, Brownian motion is *normally distributed*.

---

[8]Strictly speaking, this is only true if $tM \in \mathbf{Z}$. If not, then it is still true up to errors of size $O(1/\sqrt{M})$, which will not matter when $M \to \infty$. Thus, for convenience, we ignore those $O(1/\sqrt{M})$ errors in our discussion.

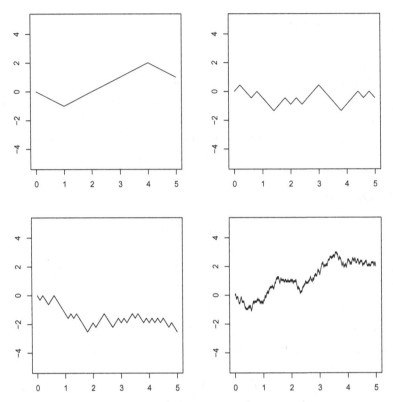

Figure 18: Simulations of the Brownian motion constructor $\{Y_t^{(M)}\}$, with $M = 1$ (top left), $M = 5$ (top right), $M = 10$ (bottom left), and $M = 100$ (bottom right).

Third, $Y_s^{(M)} - Y_t^{(M)} = \frac{1}{\sqrt{M}}(Z_{tM+1} + Z_{tM+2} + \ldots + Z_{sM})$ for $0 < t < s$. So, as $M \to \infty$, $Y_s^{(M)} - Y_t^{(M)} \to \text{Normal}(0, s - t)$, <u>independent</u> of $Y_t^{(M)}$.

So, $B_s - B_t \sim \text{Normal}(0, s - t)$, and is independent of B_t.
More generally, if $0 \leq t_1 \leq s_1 \leq t_2 \leq s_2 \leq \ldots \leq t_k \leq s_k$, then $B_{s_i} - B_{t_i} \sim \text{Normal}(0, s_i - t_i)$, and $\{B_{s_i} - B_{t_i}\}_{i=1}^k$ are all independent. That is, Brownian motion has *independent normal increments*.

Fourth, if $0 < t \leq s$, then

$$\mathbf{Cov}(B_t, B_s) = \mathbf{E}(B_t B_s) = \mathbf{E}(B_t[B_s - B_t + B_t])$$

$$= \mathbf{E}(B_t[B_s - B_t]) + \mathbf{E}((B_t)^2)$$

$$= \mathbf{E}(B_t)\,\mathbf{E}(B_s - B_t) + \mathbf{E}((B_t)^2)$$

$$= (0)(0) + t = t.$$

In general, $\mathbf{Cov}(B_t, B_s) = \min(t, s)$, which is the *covariance structure* of Brownian motion.

Finally, the processes $\{Y_t^{(M)}\}$ in Figure 18 have *continuous sample paths*, i.e. the random function $t \to Y_t^{(M)}$ is always continuous.

Inspired by all of the above, we define[9]:

(4.1.1) Definition. *Brownian motion* is a continuous-time process $\{B_t\}_{t \geq 0}$ satisfying the properties that:
1. $B_0 = 0$.
2. Normally distributed: $B_t \sim \text{Normal}(0, t)$;
3. Independent normal increments (as above);
4. Covariance structure: $\mathbf{Cov}(B_s, B_t) = \min(s, t)$;
5. Continuous sample paths (the function $t \to B_t$ is continuous).

Aside: Although the function $t \mapsto B_t$ is continuous everywhere, w.p. 1 it is differentiable nowhere, i.e. it does not have any derivatives. Intuitive reason #1: $\{Y_t^{(M)}\}$ has non-differentiable "spikes" at $t = i/M$ for all $i \in \mathbf{Z}$, which includes more and more points as $M \to \infty$. Intuitive reason #2: If it were differentiable, its derivative at t would be $\lim_{h \searrow 0} \frac{1}{h}(B_{t+h} - B_t)$. But $B_{t+h} - B_t \sim \text{Normal}(0, h)$, so $\frac{1}{h}(B_{t+h} - B_t) \sim \text{Normal}(0, 1/h)$, with variance $1/h$ which $\to \infty$ as $h \searrow 0$. Hence, the limit for the derivative does not converge as $h \searrow 0$.

(4.1.2) Problem. Compute $\mathbf{E}[(B_2 + B_3 + 1)^2]$. [sol]

(4.1.3) Problem. Compute $\mathbf{Var}[B_3 + B_5 + 6]$. [sol]

(4.1.4) Problem. Let $\{B_t\}_{t \geq 0}$ be Brownian motion. Compute $\mathbf{Var}(B_5 B_8)$, the variance of $B_5 B_8$. [Hint: You may use without

---

[9]Our derivation of Brownian motion from the $\{Y_t^{(M)}\}$ processes is just intuitive. A formal proof of the existence of Brownian motion requires measure theory; see e.g. Billingsley (1995), or Fristedt and Gray (1997), or many other advanced stochastic processes books.

proof that if $Z \sim \text{Normal}(0, 1)$, then $\mathbf{E}(Z) = \mathbf{E}(Z^3) = 0$, $\mathbf{E}(Z^2) = 1$, and $\mathbf{E}(Z^4) = 3$. Also, $B_8 = B_5 + (B_8 - B_5)$.]

(4.1.5) Example. Let $\alpha > 0$, and let $W_t = \alpha B_{t/\alpha^2}$.
Then $B_{t/\alpha^2} \sim \text{Normal}(0, t/\alpha^2)$.
Hence, $W_t \sim \text{Normal}(0, \alpha^2(t/\alpha^2)) = \text{Normal}(0, t)$.
For $0 < t < s$, $\mathbf{E}(W_t W_s) = \alpha^2 \mathbf{E}(B_{t/\alpha^2} B_{s/\alpha^2}) = \alpha^2(t/\alpha^2) = t$.
In fact, $\{W_t\}$ has all the same properties as $\{B_t\}$.
So, $\{W_t\}$ "is" Brownian motion, too.
Indeed, $\{W_t\}$ is a *transformation* of our original Brownian motion $\{B_t\}$, with just as much claim to "be" Brownian motion as $\{B_t\}$.

(4.1.6) Problem. Let $\{B_t\}_{t \geq 0}$ be Brownian motion, and let $\alpha > 0$.
Show each of the following is also Brownian motion:
(a) $\{X_t\}_{t \geq 0}$, where $X_t = B_{t+\alpha} - B_\alpha$.
(b) $\{Y_t\}_{t \geq 0}$, where $Y_t = \alpha B_{t/\alpha^2}$.

We now turn to conditional distributions, for $0 < t < s$.
Given B_r for $0 \leq r \leq t$, what is the conditional distribution of B_s?
Well, $B_s = B_t + (B_s - B_t)$.
So, given B_t, B_s equals B_t plus an independent $\text{Normal}(0, s - t)$.
That is, $B_s \mid B_t = B_t + \text{Normal}(0, s - t) \sim \text{Normal}(B_t, s - t)$.
Hence, given B_t, B_s is normal with mean B_t and variance $s - t$.
So, in particular, $\mathbf{E}[B_s \mid \{B_r\}_{0 \leq r \leq t}] = B_t$.
It follows that $\{B_t\}$ is a (continuous-time) *martingale*!
So, we can apply previous martingale results like the Optional Stopping Corollary (3.2.6) to Brownian motion, just as in discrete time.

(4.1.7) Example. Let $a, b > 0$, and let $\tau = \min\{t \geq 0 : B_t = -a \text{ or } b\}$. What is $s \equiv \mathbf{P}(B_\tau = b)$?
Well, here $\{B_t\}$ is martingale, and τ is stopping time.
Furthermore, $\{B_t\}$ is bounded up to time τ, i.e. $|B_t| \mathbf{1}_{t \leq \tau} \leq \max(|a|, |b|)$.
So, by the Optional Stopping Corollary (3.2.6), we must have $\mathbf{E}(B_\tau) = \mathbf{E}(B_0) = 0$. Hence, $s(b) + (1 - s)(-a) = 0$.
It follows that $s = \frac{a}{a+b}$ (similar to Gambler's Ruin (1.7.1) when $p = 1/2$).

In Example (4.1.7), we can ask, what is the expected time $\mathbf{E}(\tau)$ that it takes for Brownian motion to hit $-a$ or b?

To answer this question, we need one more result:

(4.1.8) Proposition. Let $Y_t = B_t^2 - t$. Then $\{Y_t\}$ is a martingale.

Proof: For $0 < t < s$, $\mathbf{E}[Y_s \,|\, \{B_r\}_{r \leq t}] = \mathbf{E}[B_s^2 - s \,|\, \{B_r\}_{r \leq t}]$
$= \mathbf{Var}[B_s \,|\, \{B_r\}_{r \leq t}] + (\mathbf{E}[B_s \,|\, \{B_r\}_{r \leq t}])^2 - s$
$= (s - t) + (B_t)^2 - s = (B_t)^2 - t = Y_t$.
Then by the Double-Expectation Formula (A.6.6),

$$\mathbf{E}[Y_s \,|\, \{Y_r\}_{r \leq t}] \;=\; \mathbf{E}\Big[\mathbf{E}[Y_s \,|\, \{B_r\}_{r \leq t}] \,\Big|\, \{Y_r\}_{r \leq t}\Big]$$

$$= \mathbf{E}\big[Y_t \,|\, \{Y_r\}_{r \leq t}\big] \;=\; Y_t . \qquad\blacksquare$$

Example (4.1.7) continued: Recall that $\tau = \min\{t \geq 0 : B_t = -a \text{ or } b\}$. Then what is $e := \mathbf{E}(\tau)$?
Let $Y_t = B_t^2 - t$. Then $\{Y_t\}$ is a martingale by Proposition (4.1.8).
But $\mathbf{E}(Y_\tau) = \mathbf{E}(B_\tau^2 - \tau) = \mathbf{E}(B_\tau^2) - \mathbf{E}(\tau) = pb^2 + (1-p)(-a)^2 - e = \frac{a}{a+b}b^2 + \frac{b}{a+b}a^2 - e = ab - e$.
We then claim that $\mathbf{E}(Y_\tau) = \mathbf{E}(Y_0) = 0$. Indeed:
> Let $\tau_M = \min(\tau, M)$. Then τ_M is bounded, so $\mathbf{E}(Y_{\tau_M}) = 0$.
> But $Y_{\tau_M} = B_{\tau_M}^2 - \tau_M$, so $\mathbf{E}(\tau_M) = \mathbf{E}(B_{\tau_M}^2)$.
> As $M \to \infty$, $\mathbf{E}(\tau_M) \to \mathbf{E}(\tau)$ by the Monotone Convergence Theorem (A.10.2), and $\mathbf{E}(B_{\tau_M}^2) \to \mathbf{E}(B_\tau^2)$ by the Bounded Convergence Theorem (A.10.1).
> Therefore, $\mathbf{E}(\tau) = \mathbf{E}(B_\tau^2)$, i.e. $\mathbf{E}(Y_\tau) = 0$ as claimed.

So, $\mathbf{E}(Y_\tau) = 0$, i.e. $ab - e = 0$, i.e. $e = ab$.
So, on average it takes time ab for Brownian motion to hit $-a$ or b, exactly like the discrete Gambler's Ruin result (3.3.4).

Finally, suppose that

(4.1.9) $$X_t \;=\; x_0 + \mu t + \sigma B_t , \quad t \geq 0 ,$$

for some constants x_0, μ, and $\sigma \geq 0$.
Then we say that $\{X_t\}$ is a *diffusion*, and call x_0 the *initial value*, and μ the *drift*, and σ the *volatility*.
The values of μ and σ affect the behaviour of the resulting diffusion; see Figure 19.
In fancier notation, we can write $X_0 = x_0$ and $dX_t = \mu\,dt + \sigma\,dB_t$.

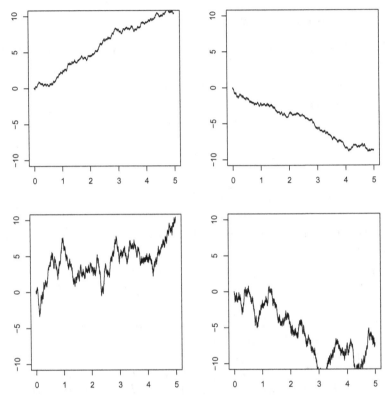

Figure 19: Diffusions with positive (left) or negative (right) drift, and small (top) or large (bottom) volatility.

We can then compute that $\mathbf{E}(X_t) = x_0 + \mu t$, and $\mathbf{Var}(X_t) = \sigma^2 t$, and $\mathbf{Cov}(X_t, X_s) = \sigma^2 \min(s, t)$, and $X_t \sim N(x_0 + \mu t, \sigma^2 t)$.

(4.1.10) Example. Suppose $X_t = 2 + 5t + 3B_t$ for $t \geq 0$.
What are $\mathbf{E}(X_t)$ and $\mathbf{Var}(X_t)$ and $\mathbf{Cov}(X_t, X_s)$?
Well, $\mathbf{E}(X_t) = 2 + 5t$, and $\mathbf{Var}(X_t) = 3^2 \mathbf{Var}(B_t) = 9t$.
Also for $0 < t < s$, $\mathbf{Cov}(X_t, X_s) = \mathbf{E}[(X_t - 5t - 2)(X_s - 5s - 2)] = \mathbf{E}[(3B_t)(3B_s)] = 9\,\mathbf{E}[B_t\,B_s] = 9t$.
It also follows that $X_t \sim \text{Normal}(2 + 5t, 9t)$.
In fancier notation: $X_0 = 2$ and $dX_t = 5\,dt + 3\,dB_t$.

Aside: More generally, we could have a diffusion with $dX_t =$

$\mu(X_t)\,dt + \sigma(X_t)\,dB_t$, where μ and σ are <u>functions</u> of the current value X_t. Such processes are more complicated, and we do not pursue them here; for discussion see any advanced book on diffusions.

(4.1.11) Problem. Let $\{B_t\}_{t\geq 0}$ be Brownian motion, and let $Y_t = 2 + 3t + 4B_t$ be a diffusion process. Compute $\mathbf{E}(Y_3 Y_5)$.

(4.1.12) Problem. Let $\{B_t\}_{t\geq 0}$ be Brownian motion.
(a) Compute $\mathbf{Cov}(B_6, B_8)$, i.e. the covariance of B_6 and B_8.
(b) Compute $\mathbf{Corr}(B_6, B_8)$, i.e. the correlation of B_6 and B_8.
(c) Compute $\lim_{h\searrow 0} \mathbf{E}\left(\frac{B_{8+h}-B_8}{h}\right)$.
(d) Compute $\lim_{h\searrow 0} \mathbf{E}\left(\frac{(B_{8+h}-B_8)^2}{h}\right)$.

(4.1.13) Problem. Let $\{B_t\}_{t\geq 0}$ be Brownian motion. Let $\theta \in \mathbf{R}$, and let $Y_t = \exp(\theta B_t - \theta^2 t/2)$. Prove that $\{Y_t\}_{t\geq 0}$ is a martingale. [Hint: You may use without proof that if $Z \sim \text{Normal}(0,1)$, and $a \in \mathbf{R}$, then $\mathbf{E}[e^{aZ}] = e^{a^2/2}$.]

4.2. Application – Stock Options (Continuous)

Assume now that a stock price X_t is equal to some (random) positive value, for each <u>continuous</u> time $t \geq 0$.

A common model is to assume that

$$(4.2.1) \qquad\qquad X_t = x_0 \exp(\mu t + \sigma B_t).$$

That is, changes occur <u>proportional</u> to the total price.
(That makes sense; price proportions are what is important.)
If $Y_t = \log(X_t)$, then $Y_t = y_0 + \mu t + \sigma B_t$, i.e. $dY_t = \mu\,dt + \sigma\,dB_t$.
So, $\{Y_t\} = \{\log(X_t)\}$ is a *diffusion* as in (4.1.9), with *drift* (or *appreciation rate*) μ, and *volatility* σ.

We also assume a *risk-free interest rate* r, so that \$1 at time 0 (i.e. "today") is worth \$$e^{rt}$ at a time t years later.
Equivalently, \$1 at a future time $t > 0$ is worth \$$e^{-rt}$ at time 0.
So, the "discounted" stock price (in "today's dollars") is

$$D_t \equiv e^{-rt}X_t = e^{-rt}x_0\exp(\mu t + \sigma B_t) = x_0\exp((\mu - r)t + \sigma B_t).$$

As a special case, if $r = 0$, then there is no discounting.
(Discounting might not seem important, but it turns out to provide
a key step below.)
Then, just like in the discrete case (Section 3.7), we define:

(4.2.2) Definition. A *(European call) stock option* is the option
to buy one share of a stock for a fixed *strike price* (or, *exercise price*)
K at a fixed future *strike time* (or, *maturity time*) $S > 0$.

So what is such an option worth?
Well, at time S, it is worth $\max(0, X_S - K)$ (since if $X_S < K$ you
won't exercise the option, but if $X_S > K$ you can buy it for \$$K$ and
then sell it for \$$X_S$ for a net profit of $X_S - K$).
So, at time 0, its discounted value is $e^{-rS} \max(0, X_S - K)$.
But at time 0, X_S is unknown (random), so this time 0 discounted
value is a random variable and thus more challenging.

The question is: What is the "fair price" of this option?
As before, this means the "no-arbitrage" price, i.e. a price such that
you cannot make a guaranteed profit by combinations of buying or
selling the option, and buying and selling the stock.
Note: As before, we assume the ability to buy or sell arbitrary
amounts of the stock and/or the option at any time, infinitely of-
ten, including going negative (i.e., "shorting" the stock or option),
with no transaction fees.
So, what is the fair price at time 0?
Is it the discounted value $e^{-rS} \max(0, X_S - K)$?
No, that value is unknown, so it cannot be used as a price.
Is it simply the expected value, $\mathbf{E}[e^{-rS} \max(0, X_S - K)]$?
No, this would allow for arbitrage!

How, then, can we compute the fair option price?
We use the Martingale Pricing Principle (3.7.4) again!
Key: If $\mu = r - \frac{\sigma^2}{2}$, then it can be computed (see Problem (4.2.7))
that $\{D_t\}$ becomes a martingale.
And then, the option values become[10] a martingale too!

---

[10]For a proof, see e.g. Theorem 1.2.1 of I. Karatzas (1997), *Lectures on the
Mathematics of Finance*, CRM Monograph Series **8**, American Mathematical Society, or many other books on mathematical finance.

Hence, as in the discrete case, the Martingale Pricing Principle holds: The fair, no-arbitrage price of the option at time zero is the same as the <u>expected value</u> of the option at time S under the martingale probabilities where $\mu = r - \frac{\sigma^2}{2}$. That is:

(4.2.3) Proposition. The fair price for the above option is equal to $\mathbf{E}[e^{-rS}\max(0, X_S - K)]$, but only after <u>replacing</u> μ by $r - \frac{\sigma^2}{2}$, i.e. such that $X_S = x_0 \exp([r - \frac{\sigma^2}{2}]S + \sigma B_S)$.

So, the fair price is an expected value of a function of B_S. But we know that $B_S \sim \text{Normal}(0, S)$. So, the fair price can be computed by an <u>integral</u> with respect to the normal density of B_S. After some computations (Problem (4.2.8)), we find:

(4.2.4) Black-Scholes Formula. The fair (no-arbitrage) price at time 0 of an option to buy one share of a stock governed by (4.2.1) at maturity time S for exercise price $\$K$ is equal to:

$$x_0 \; \Phi\left(\frac{(r + \frac{\sigma^2}{2})S - \log(K/x_0)}{\sigma\sqrt{S}}\right)$$

$$- e^{-rS} K \; \Phi\left(\frac{(r - \frac{\sigma^2}{2})S - \log(K/x_0)}{\sigma\sqrt{S}}\right),$$

where $\Phi(u) = \int_{-\infty}^{u} \frac{1}{\sqrt{2\pi}} e^{-x^2/2}\, dx$ as in (A.3.12).

This formula originally appeared as equation (13) of F. Black and M. Scholes (1973), "The Pricing of Options and Corporate Liabilities", *Journal of Political Economy* **81(3)**, 637–654.
Despite its unrealistic assumptions (e.g. no transaction fees), it is very widely used to price actual stock options. Indeed, many web pages[11] provide online calculators to compute its values.

Aside: The price (4.2.4) does <u>not</u> depend on the appreciation rate μ, since we first have to <u>replace</u> μ by $r - \frac{\sigma^2}{2}$. This seems surprising! How could it be?

---

[11]e.g. `https://www.mystockoptions.com/black-scholes.cfm`

Well, intuitively, if μ is large, then we can already make a large profit by buying the stock itself, so the option doesn't add much in the way of additional value (despite the large μ).

However, the price is an increasing function of the volatility σ. This makes sense, since the option "protects" you against potential large drops in the stock price.

(4.2.5) Example. Suppose $x_0 = 100$ (dollars), $K = 110$ (dollars), $S = 1$ (years), $r = 0.05$ (i.e. 5% per year), and $\sigma = 0.3$ (i.e. 30% per year). Then the fair option price is computed to be: $10.02.
Or, if $x_0 = 200$, $K = 250$, $S = 2$, $r = 0.1$, and $\sigma = 0.5$, then the price becomes: $53.60.
Or, if $x_0 = 200$, $K = 250$, $S = 2$, and $r = 0.1$, but $\sigma = 0.8$, then the price becomes: $84.36. This value is higher, which makes sense since the volatility is larger.

(4.2.6) Problem. Compute the fair price of an option where $x_0 = 40$, $K = 30$, $S = 3$, $r = 0$, and either **(i)** $\sigma = 0.1$, or **(ii)** $\sigma = 0.3$. Which is larger, and why? **[sol]**

(4.2.7) Problem. Let $\{B_t\}_{t\geq 0}$ be Brownian motion, let $X_t = x_0 \exp(\mu t + \sigma B_t)$ be the stock price model (where $\sigma > 0$), and let $D_t = e^{-rt} X_t$ be the discounted stock price. Show that if $\mu = r - \frac{\sigma^2}{2}$, then $\{D_t\}$ is a martingale. [Hint: Don't forget Problem (4.1.13).]

(4.2.8) Problem. Let $\{B_t\}_{t\geq 0}$ be Brownian motion, and let $X_t = x_0 \exp(\mu t + \sigma B_t)$ be the stock price model (where $\sigma > 0$). Show that if $\mu = r - \frac{\sigma^2}{2}$, then $\mathbf{E}\left[e^{-rS} \max(0, X_S - K)\right]$ is equal to the formula (4.2.4). [Hint: Write the expectation as an integral with respect to the density function for B_S. Then, break up the integral into the part where $X_S - K \geq 0$ and the part where $X_S - K < 0$.]

(4.2.9) Problem. Consider the price formula (4.2.4), with r, σ, S, and x_0 fixed positive quantities.
(a) What happens to the price (4.2.4) as $K \searrow 0$? Does this result make intuitive sense?
(b) What happens to the price (4.2.4) as $K \to \infty$? Does this result make intuitive sense?

(4.2.10) Problem. Consider the price formula (4.2.4), with r, σ, x_0, and K fixed positive quantities.
(a) What happens to the price (4.2.4) as $S \searrow 0$? [Hint: Consider separately the cases $K > x_0$, $K = x_0$, and $K < x_0$.] Does this make intuitive sense?
(b) What happens to the price (4.2.4) as $T \to \infty$? Does this result make intuitive sense?

4.3. Poisson Processes

We begin with a Motivating Example:
Suppose your city has an average of $\lambda = 2.5$ house fires per day.
Intuitively, this is caused by a very <u>large</u> number n of buildings, each of which has a very <u>small</u> probability p of having a fire.
If so, then the mean number of fires is np.
Hence, we must have $np = \lambda$, so $p = \lambda/n$.
Then the number of fires today has the distribution Binomial$(n, p) = $ Binomial$(n, \lambda/n)$.
By the binomial formula (A.3.2), this means that

$$\mathbf{P}(\#\text{fires} = k) = \binom{n}{k} p^k (1-p)^{n-k}$$

$$= \frac{n(n-1)(n-2)\ldots(n-k+1)}{k!} (\lambda/n)^k (1-(\lambda/n))^{n-k}.$$

Now, as $n \to \infty$ with k fixed, it follows using (A.4.9) that

$$(4.3.1) \qquad \mathbf{P}(\#\text{fires} = k) \to \frac{n^k}{k!} (\lambda/n)^k (e^{-\lambda/n})^n = \frac{1}{k!} \lambda^k e^{-\lambda}.$$

This is equal to the Poisson(λ) distribution (A.3.5).
We conclude that the number of fires in a given day is well modelled by the distribution Poisson(λ).
So, if $\lambda = 2.5$, then # fires today $\sim$ Poisson(2.5), and $\mathbf{P}(\#$ fires today $= k) \approx e^{-2.5} \frac{(2.5)^k}{k!}$, for $k = 0, 1, 2, 3, \ldots$
Similarly, # fires tomorrow $\sim$ Poisson(2.5).

Then, assuming that today and tomorrow are independent, the Poisson additivity property (A.3.7) says that the number of fires today and tomorrow combined $\sim$ Poisson$(2 * \lambda) =$ Poisson(5), etc. This model of house fires corresponds to a *Poisson process*.

To construct a Poisson process requires several ingredients:

Let $\{Y_n\}_{n=1}^{\infty}$ be i.i.d. $\sim$ Exponential(λ), for some *rate* $\lambda > 0$. So, by (A.3.10), Y_n has density function $\lambda e^{-\lambda y}$ for $y > 0$, and $\mathbf{P}(Y_n > y) = e^{-\lambda y}$ for $y > 0$, and $\mathbf{E}(Y_n) = 1/\lambda$.

Then, let $T_0 = 0$, and $T_n = Y_1 + Y_2 + \ldots + Y_n$ for $n \geq 1$. Then T_n is the time of the nth *event* (or *arrival*). In the Motivating Example, T_n would be the time of the nth fire.

Next, let $N(t) = \max\{n \geq 0 : T_n \leq t\} = \#\{n \geq 1 : T_n \leq t\}$. That is, $N(t)$ is the number of events (or *arrivals*) up to time t. Here $\{N(t)\}$ is a *Poisson process* with *intensity* λ. It is sometimes called a *counting process*, since it "counts" the number of events or arrivals up to time t. In our Motivating Example, $N(t)$ would be the number of fires between times 0 and t.

Then we can ask, what is distribution of $N(t)$, i.e. $\mathbf{P}(N(t) = k)$?

(4.3.2) Proposition.   We have that $N(t) \sim$ Poisson(λt), i.e. $\mathbf{P}(N(t) = k) = \frac{(\lambda t)^k}{k!} e^{-\lambda t}$, for $k = 0, 1, 2, 3, \ldots$. (So, $\mathbf{E}[N(t)] = \mathbf{Var}[N(t)] = \lambda t$.)

Proof: Well, $N(t) = k$ iff both $T_k \leq t$ and $T_{k+1} > t$, i.e. iff there is $0 \leq s \leq t$ with $T_k = s$ and $T_{k+1} - T_k > t - s$. So,

$$\mathbf{P}(N(t) = k) \;=\; \mathbf{P}(T_k \leq t, \; T_{k+1} > t)$$

$$= \; \mathbf{P}(\exists\, 0 \leq s \leq t : \; T_k = s, \; Y_{k+1} > t - s)$$

$$= \; \int_0^t f_{T_k}(s) \, \mathbf{P}(Y_{k+1} > t - s) \, ds$$

$$= \; \int_0^t f_{T_k}(s) \, e^{-\lambda(t-s)} \, ds \, ,$$

where f_{T_k} is the density function for T_k. It remains to compute f_{T_k}, and hence compute this integral.

To do so, recall that $Y_n \sim$ Exponential$(\lambda) = $ Gamma$(1, \lambda)$, a Gamma distribution, so $T_k := Y_1 + Y_2 + \ldots + Y_k \sim$ Gamma(k, λ), with density function $f_{T_k}(s) = \frac{\lambda^k}{\Gamma(k)} s^{k-1} e^{-\lambda s} = \frac{\lambda^k}{(k-1)!} s^{k-1} e^{-\lambda s}$. Hence,

$$\mathbf{P}(N(t) = k) = \int_0^t \frac{\lambda^k}{(k-1)!} s^{k-1} e^{-\lambda s} e^{-\lambda(t-s)} \, ds$$

$$= \frac{\lambda^k}{(k-1)!} e^{-\lambda t} \int_0^t s^{k-1} \, ds = \frac{\lambda^m}{(k-1)!} e^{-\lambda t} [\frac{t^k}{k}] = \frac{(\lambda t)^k}{k!} e^{-\lambda t},$$

as claimed.   ∎

Now, recall the *memoryless property* (A.6.9) of the Exponential distribution.

This implies that the $\{Y_n\}$, and hence also the process $\{N(t)\}$, "starts over again", completely freshly and independently, at each new time s. It follows that $N(t + s) - N(s)$ has the same distribution as $N(t)$, i.e. that $N(t + s) - N(s) \sim N(t) \sim$ Poisson(λt) for all $s, t > 0$. It also follows that if $0 \le a < b \le c < d$, then $N(d) - N(c)$ is independent of $N(b) - N(a)$, and similarly for multiple non-overlapping time intervals.

More generally, if $0 \le t_1 \le s_1 \le t_2 \le s_2 \le \ldots \le t_k \le s_k$, then $N(s_i) - N(t_i) \sim$ Poisson$(\lambda(s_i - t_i))$, and $\{N(s_i) - N(t_i)\}_{i=1}^k$ are all independent, i.e. it has *independent Poisson increments*.

Hence, our formal definition of a Poisson Process is:

(4.3.3) Definition.   A *Poisson process* with intensity $\lambda > 0$ is a collection $\{N(t)\}_{t \ge 0}$ of non-decreasing integer-valued random variables satisfying the properties that:

1. $N(0) = 0$;
2. $N(t) \sim$ Poisson(λt) for all $t \ge 0$; and
3. Independent Poisson increments, as above.

Consider again our Motivating Example, where there is an average of 2.5 fires per day, and the number of fires approximately follows a Poisson process with intensity $\lambda = 2.5$.

Then, as above, the number of fires today and tomorrow combined has distribution Poisson$(2 * 2.5) = $ Poisson(5), so e.g. $\mathbf{P}(9$ fires today and tomorrow combined$) = e^{-2*2.5} \frac{(2*2.5)^9}{9!} = e^{-5}(\frac{5^9}{9!}) \doteq 0.036$, etc.

Similarly, the number of fires in the next <u>hour</u> has distribution given by Poisson(2.5/24), so e.g. $\mathbf{P}$(at least one fire in next hour) $= 1 - \mathbf{P}$(no fires in next hour) $= 1 - \mathbf{P}(N(1/24) = 0) = 1 - e^{-2.5/24}\frac{(2.5/24)^0}{0!} \doteq 1 - 0.90 = 0.10$. And, $\mathbf{P}$(exactly 3 fires in next hour) $= e^{-2.5/24}\frac{(2.5/24)^3}{3!} \doteq 0.00017 \doteq 1/5891$, etc.

Joint probabilities can be converted to non-overlapping intervals:

(4.3.4) Example. If $\{N(t)\}$ is a Poisson process with $\lambda = 2$, then

$$\mathbf{P}[N(3) = 5, \ N(3.5) = 9] = \mathbf{P}[N(3) = 5, \ N(3.5) - N(3) = 4]$$

$$= \mathbf{P}[N(3) = 5] \ \mathbf{P}[N(3.5) - N(3) = 4]$$

$$= \left[e^{-\lambda(3)}\frac{(\lambda(3))^5}{5!} \right] \left[e^{-\lambda(0.5)}\frac{(\lambda(0.5))^4}{4!} \right]$$

$$= \left(e^{-6}\frac{6^5}{120} \right) \left(e^{-1}\frac{1^4}{24} \right) = e^{-7}(2.7) \doteq 0.0025 \doteq 1/400 .$$

Conditional probabilities can also provide new insights into the distributions of the Poisson process events:

(4.3.5) Example. Let $\{N(t)\}$ be a Poisson process with intensity λ. Then for $0 < t < s$,

$$\mathbf{P}(N(t) = 1 \mid N(s) = 1) = \frac{\mathbf{P}(N(t) = 1, \ N(s) = 1)}{\mathbf{P}(N(s) = 1)}$$

$$= \frac{\mathbf{P}(N(t) = 1, \ N(s) - N(t) = 0)}{\mathbf{P}(N(s) = 1)}$$

$$= \frac{e^{-\lambda t}\frac{(\lambda t)^1}{1!} e^{-\lambda(s-t)}\frac{(\lambda(s-t))^0}{0!}}{e^{-\lambda s}\frac{(\lambda s)^1}{1!}} = t/s .$$

That is, conditional on having just $N(s) = 1$ event by time s, the conditional distribution of that one event is <u>uniform</u> over the interval $[0, s]$ (and does not depend on the value of λ).

Similarly, we compute that e.g.

$$\mathbf{P}(N(4) = 1 \mid N(5) = 3) = \frac{\mathbf{P}(N(4) = 1, \ N(5) = 3)}{\mathbf{P}(N(5) = 3)}$$

$$= \frac{\mathbf{P}(N(4) = 1, \ N(5) - N(4) = 2)}{\mathbf{P}(N(5) = 3)}$$

$$= \frac{(e^{-4\lambda}(4\lambda)^1/1!)(e^{-\lambda}\lambda^2/2!)}{e^{-5\lambda}(5\lambda)^3/3!} = \frac{(4)^1/1!)(1/2!)}{(5)^3/3!}$$

$$= \frac{4/2}{125/6} = 24/250 = 12/125.$$

This also does not depend on λ.

Furthermore, it equals $\binom{3}{1}(4/5)^1(1/5)^2$, which is the probability that a Binomial$(3, 4/5)$ random variable (A.3.2) equals 1.

This says that, conditional on having $N(5) = 3$ events by time 5, the number of events by time 4 is Binomial$(3, 4/5)$.

That is, conditional on $N(5) = 3$, those three events are i.i.d. $\sim$ Uniform$[0, 5]$.

(4.3.6) Problem. Let $\{N(t)\}_{t\geq 0}$ be a Poisson process with $\lambda = 4$.
(a) Compute $\mathbf{P}[N(3) = 5]$. [sol]
(b) Compute $\mathbf{P}[N(3) = 5 \mid N(2) = 1]$. [sol]
(c) Compute $\mathbf{P}[N(3) = 5 \mid N(2) = 1, \ N(6) = 6]$. [sol]

(4.3.7) Problem. Let $\{N(t)\}_{t\geq 0}$ be a Poisson process with rate $\lambda = 3$, and arrival times $T_1, T_2, T_3, \ldots$.
(a) Let $X = N(8) - N(5)$ be the number of arrivals between times 5 and 8. Compute $\mathbf{E}(X)$ and $\mathbf{P}(X = 0)$.
(b) Let $Y = \inf\{T_i : T_i > 5\}$ be the first arrival after time 5. Compute $\mathbf{E}(Y)$.

(4.3.8) Problem. Let $\{N(t)\}_{t\geq 0}$ be a Poisson process with intensity $\lambda = 5$. Let $X = N(6) - N(1)$, and let $Y = N(5) - N(3)$. Compute $\mathbf{E}[XY]$. [Hint: Don't forget (A.3.6).]

Alternate Characterisations:

There are other ways to characterise Poisson processes.

One makes use of *order notation* $o(\cdot)$ (cf. Section A.4):

(4.3.9) Proposition. If $N(t)$ is a Poisson process with rate $\lambda > 0$, then as $h \searrow 0$:

(a) $\mathbf{P}(N(t+h) - N(t) = 1) = \lambda h + o(h)$.
(b) $\mathbf{P}(N(t+h) - N(t) \geq 2) = o(h)$.

Proof: Here $\mathbf{P}(N(t+h) - N(t) = 1) = \mathbf{P}(N(h) = 1) = e^{\lambda h}(\lambda h)^1/1! = [1 + \lambda h + O(h^2)](\lambda h) = \lambda h + \lambda^2 h^2 + O(h^3) = \lambda h + O(h^2) = \lambda h + o(h)$.

Also, $\mathbf{P}(N(t+h) - N(t) \geq 2) = \mathbf{P}(N(h) \geq 2) = 1 - \mathbf{P}(N(h) = 0) - \mathbf{P}(N(h) = 1) = 1 - e^{\lambda h}(\lambda h)^0/0! - e^{\lambda h}(\lambda h)^1/1! = 1 - [1 + \lambda h + O(h^2)](1) - [1 + \lambda h + O(h^2)](\lambda h) = 1 - 1 + \lambda h - \lambda h + \lambda^2 h^2 + O(h^2) + O(h^3) = O(h^2) = o(h)$. ∎

In fact, Proposition (4.3.9) has a converse: <u>Any</u> stochastic process with independent increments which satisfies properties (a) and (b) must be a Poisson process with rate λ.
This provides a second way to define Poisson processes.

A third way is that Poisson processes are the <u>only</u> integer-valued processes with independent increments (not necessarily Poisson) such that $\mathbf{E}[N(s+t) - N(s)] = \lambda t$ for all $s, t > 0$.

<u>Aside:</u> There are also *time-inhomogeneous* Poisson processes, where $\lambda = \lambda(t)$, and $N(b) - N(a) \sim \text{Poisson}\left(\int_a^b \lambda(t)\,dt\right)$. It is also possible to have Poisson processes on other <u>regions</u>, e.g. in multiple dimensions, etc.; for a two-dimensional illustration, see: www.probability.ca/pois

Poisson Clumping:

Suppose we have a Poisson Process on $[0,100]$ with intensity $\lambda = 1$ event per day. So, we <u>expect</u> to see approximately one event per day. What is $\mathbf{P}(\exists r \in [1, 100] : N(r) - N(r-1) = 4)$, the probability of seeing 4 events within some one single day? Well,

$$\mathbf{P}(\exists r \in [1, 100] : N(r) - N(r-1) = 4)$$

$$\geq \ \mathbf{P}(\exists m \in \{1, 2, \ldots, 100\} : N(m) - N(m-1) = 4)$$

$$= \ 1 - \mathbf{P}(\nexists m \in \{1, 2, \ldots, 100\} : N(m) - N(m-1) = 4)$$

$$= \ 1 - (\mathbf{P}(N(1) \neq 4))^{100} \ = \ 1 - (1 - P(N(1) = 4))^{100}$$

$$= \ 1 - (1 - e^{-1}(1^4/4!))^{100} \ = \ 1 - (1 - 1/24e)^{100} \ \doteq \ 0.787 \,.$$

That is, there is nearly an 80% chance of some such clumps!

This is an illustration of *Poisson clumping*, whereby the $\{T_i\}$ tend to "clump up" in various patterns, just by chance alone. This doesn't "mean" anything at all: they are still <u>independent</u>. But might "seem" like it does have meaning:

(4.3.10) Pedestrian Deaths example. (A true story.)
In January 2010, there were 7 pedestrian deaths in Toronto.
This led to lots of media hype, concerned friends, etc.
It turned out that Toronto averages about 31.9 pedestrian deaths per year, i.e. $\lambda = 2.66$ per month.
So, if it's Poisson, $\mathbf{P}(7 \text{ or more}) = \sum_{j=7}^{\infty} e^{-2.66} \frac{(2.66)^j}{j!} \doteq 1.9\%$.
This should occur about once per 52 months, or 4.4 years.
So, it's not so surprising, and doesn't indicate any change.
I told the media it was Poisson clumping, and wouldn't continue.
See e.g. the newspaper article at: www.probability.ca/ped
This was confirmed in the following 1.5 months, when there were just 2 pedestrian deaths, <u>less</u> than the expected 4.
But there were no newspaper articles written about that!

(4.3.11) Waiting Time Paradox.
Suppose there are an average of λ buses per hour. (e.g. $\lambda = 5$)
You arrive at the bus stop at a random time.
What is your expected *waiting time* until the next bus?

If the buses are completely <u>regular</u>, then there is one bus every $1/\lambda$ hours. (e.g. if $\lambda = 5$, then one bus every 12 minutes)
Then your waiting time is $\sim$ Uniform$[0, \frac{1}{\lambda}]$.
So, your mean wait is $\frac{1}{2\lambda}$ hours.
(e.g. if $\lambda = 5$, the mean is $\frac{1}{10}$ hours $= 6$ minutes)

However, suppose the buses are instead <u>completely random</u>.
Model this as $T_1, T_2, \ldots, T_n \sim$ Uniform$[0, \ n/\lambda]$, i.i.d.
Then for $0 < a < b$, as $n \to \infty$, similar to (4.3.1),

$$\#\{i : T_i \in [a, b]\} \sim \text{Binomial}\left(n, \frac{b-a}{n/\lambda}\right)$$

$$= \text{Binomial}\left(n, \frac{\lambda(b-a)}{n}\right) \to \text{Poisson}(\lambda(b-a)).$$

So, the buses form a Poisson process with rate λ.
By the memoryless property (A.6.9), your waiting time is $\sim$ Exponential(λ).
So, the mean waiting time is $\frac{1}{\lambda}$ hours. This is twice as long!
(e.g. $\lambda = 5$, mean $= \frac{1}{5}$ hours $= 12$ minutes)
But the number of buses remains the same! Is this a contradiction?
No. You are more likely to <u>arrive</u> during a longer gap.
So, your sampling of the bus gaps creates a <u>bias</u>.
That is why it is better if buses stick to their schedule!

Aside: What about <u>streetcars</u>? They can't <u>pass</u> each other, so they sometimes clump up even <u>more</u> than (independent) buses do; see e.g. the paper at: www.probability.ca/streetcarpaper

Independent Poisson processes are easily combined:

(4.3.12) Superposition. Suppose $\{N_1(t)\}_{t\geq 0}$ and $\{N_2(t)\}_{t\geq 0}$ are independent Poisson processes, with rates λ_1 and λ_2 respectively. Let $N(t) = N_1(t) + N_2(t)$. Then $\{N(t)\}_{t\geq 0}$ is also a Poisson process, with rate $\lambda_1 + \lambda_2$.

Proof: This follows since the sum of independent Poissons is always Poisson (A.3.7), and also sums of independent collections are still independent.   ∎

(4.3.13) Example. Suppose undergraduate students arrive for office hours according to a Poisson process with rate $\lambda_1 = 5$ (i.e. one every 12 minutes on average).
And, graduate students arrive independently according to a Poisson process with rate $\lambda_2 = 3$ (i.e. one every 20 minutes on average).
What is the expected time until the <u>first</u> student arrives?
Well, by Superposition (4.3.12), the total number of arrivals $N(t)$ is a Poisson process with rate $\lambda = \lambda_1 + \lambda_2 = 5 + 3 = 8$.
Let A be the time of first arrival.
Then $\mathbf{P}(A > t) = \mathbf{P}(N(t) = 0) = e^{-\lambda t}\frac{(\lambda t)^0}{0!} = e^{-\lambda t}$.
That is, $A \sim$ Exponential(λ) = Exponential(8).
Hence, $\mathbf{E}(A) = 1/\lambda = 1/8$ hours, i.e. 7.5 minutes.

Separating out the different parts of a single Poisson process is more subtle, but still possible:

(4.3.14) Thinning. Let $\{N(t)\}_{t\geq 0}$ be a Poisson process with rate λ. Suppose each arrival is independently of "type i" with probability p_i, for $i = 1, 2, 3, \ldots$, where $\sum_i p_i = 1$. (e.g. male or female, undergraduate or graduate, car or van or truck, etc.)
Let $N_i(t)$ be number of arrivals of type i up to time t.
Then the different $\{N_i(t)\}$ are all <u>independent</u>, and furthermore each $\{N_i(t)\}$ is itself a Poisson process, with rate λp_i.

Proof: Independent increments follows since arrivals for each $\{N_i(t)\}$ over non-overlapping intervals are generated from arrivals for $\{N(t)\}$ over non-overlapping intervals, which are independent.
For the independence and Poisson distributions, suppose for notational simplicity that there are just <u>two</u> types, with $p_1 + p_2 = 1$.
We need to show that:

$$\mathbf{P}(N_1(t) = j,\ N_2(t) = k) = \left(e^{-(\lambda p_1 t)}(\lambda p_1 t)^j/j!\right)\left(e^{-(\lambda p_2 t)}(\lambda p_2 t)^k/k!\right).$$

Combining the Poisson and Binomial distributions, we compute that:
$\mathbf{P}(N_1(t) = j,\ N_2(t) = k) = \mathbf{P}(j + k$ arrivals up to time t, of which j are of type 1 and k are of type 2)
$= \left(e^{-\lambda t}(\lambda t)^{j+k}/(j+k)!\right) \times \left(\binom{j+k}{j}(p_1)^j(p_2)^k\right)$
$= \left(e^{-\lambda(p_1+p_2)t}(\lambda t)^j(\lambda t)^k/(j+k)!\right) \times \left((j+k)!/j!\,k!\right)(p_1)^j(p_2)^k$
$= \left(e^{-(\lambda p_1 t)}(\lambda p_1 t)^j/j!\right)\left(e^{-(\lambda p_2 t)}(\lambda p_2 t)^k/k!\right)$, as required.   ∎

(4.3.15) Example.   Suppose students arrive for office hours according to a Poisson process with rate λ.
Suppose each student is independently either an undergraduate with probability p_1, or a graduate with probability p_2, where $p_1 + p_2 = 1$.
Then, by Thinning (4.3.14), the number of undergraduates and the number of graduates are <u>independent</u>.
Furthermore, they are each Poisson processes, with rates λp_1 and λp_2, respectively.

(4.3.16) Problem. Suppose cars arrive according to a Poisson process with rate $\lambda = 3$ cars per minute, and each car is independently either Blue with probability $1/2$, or Green with probability $1/3$, or

Red with probability $1/6$. Let $N(t)$ be the total number of cars that arrive by time t, and $N_R(t)$ the number of Red cars by time t.

(a) Compute $\mathbf{E}[(2 + N(2) + N(3))^2]$.

(b) Compute $\mathbf{E}[(N(5) - N(1))(N(7) - N(4))]$.

(c) Compute $\mathbf{P}[N_R(3) = 2, \ N(4) = 4]$.

(d) Compute $\mathbf{P}[N_R(3) = 2 \mid N(4) = 4]$.

(4.3.17) Problem. Suppose messages arrive according to a Poisson process with constant intensity $\lambda = 6$. Suppose each message is independently labelled "Urgent" with probability $1/3$. Let $M(t)$ be the total number of messages up to time t, and let $U(t)$ be the number of "Urgent" messages up to time t.

(a) Compute $\mathbf{P}[M(2) = 4]$.

(b) Compute $\mathbf{P}[U(2) = 3]$.

(c) Compute $\mathbf{P}[U(2) = 3, M(2) = 4]$.

(d) Compute $\mathbf{P}[U(2) = 3, M(5) = 4]$.

(4.3.18) Problem. Suppose buses arrive according to a Poisson process with intensity $\lambda = 3$. Suppose each bus is independently either Red with probability $1/2$, or Blue with probability $1/3$, or Green with probability $1/6$. Let T be the first positive time at which a Red bus arrives, and let U be the first positive time at which a Green bus arrives. Let $Z = \min(T, U)$.

(a) Compute $\mathbf{E}(U)$.

(b) Compute $\mathbf{P}(Z > x)$ for all $x \geq 0$.

(c) Compute $\mathbf{E}(Z)$.

4.4. Continuous-Time, Discrete-Space Processes

For most of this book, we have discussed Markov chains $\{X_n\}_{n=0}^\infty$ defined in discrete (integer) time.

But Brownian motion $\{B_t\}_{t \geq 0}$, and Poisson processes $\{N(t)\}_{t \geq 0}$, were both defined in continuous (real) time.

Can we define Markov processes in continuous time? Yes!

They are mostly just like discrete-time chains, except we need to keep track of elapsed time t too.

We again need a non-empty finite or countable state space S, and initial probabilities $\{\nu_i\}_{i \in S}$ with $\nu_i \geq 0$, and $\sum_{i \in S} \nu_i = 1$.

We also need *transition probabilities* $\{p_{ij}^{(t)}\}_{i,j \in S, t \geq 0}$ with $p_{ij}^{(t)} \geq 0$, and $\sum_{j \in S} p_{ij}^{(t)} = 1$, like before except now defined for <u>each</u> time $t \geq 0$. Then, just like the discrete definition (1.2.2), we define:

(4.4.1) Definition. A continuous-time (time-homogeneous, non-explosive) *Markov process*, on a countable (discrete) state space S, is a collection $\{X(t)\}_{t \geq 0}$ of random variables such that

$$\mathbf{P}(X_0 = i_0, X_{t_1} = i_1, X_{t_2} = i_2, \ldots, X_{t_n} = i_n)$$

$$= \nu_{i_0} p_{i_0 i_1}^{(t_1)} p_{i_1 i_2}^{(t_2 - t_1)} \cdots p_{i_{n-1} i_n}^{(t_n - t_{n-1})}$$

for all $i_0, i_1, \ldots, i_n \in S$ and all $0 < t_1 < t_2 < \ldots < t_n$.

Many properties are similar to discrete chains.
For example, $p_{ij}^{(0)} = \delta_{ij} = \begin{cases} 1, & i = j \\ 0, & i \neq j \end{cases}$
And, we can again let $P^{(t)} = \big(p_{ij}^{(t)} \big)_{i,j \in S}$ be the matrix version.
Then $P^{(0)} = I = $ identity matrix.
Also $p_{ij}^{(s+t)} = \sum_{k \in S} p_{ik}^{(s)} p_{kj}^{(t)}$, i.e. $P^{(s+t)} = P^{(s)} P^{(t)}$, which are the *Chapman-Kolmogorov equations* just like for discrete time.
If $\mu_i^{(t)} = \mathbf{P}(X(t) = i)$, and $\mu^{(t)} = \big(\mu_i^{(t)} \big)_{i \in S} = $ row vector, and $\nu = (\nu_i)_{i \in S} = $ row vector, then as before, $\mu_j^{(t)} = \sum_{i \in S} \nu_i p_{ij}^{(t)}$, and $\mu^{(t)} = \nu P^{(t)}$, and $\mu^{(t)} P^{(s)} = \mu^{(t+s)}$, etc.

However, some new issues arise in continuous time.
For example, is the mapping $t \to p_{ij}(t)$ <u>continuous</u>?
In particular, does $\lim_{t \searrow 0} p_{ij}^{(t)} = p_{ij}^{(0)} = \delta_{ij}$?
This property is called being a *standard Markov process*.
However, it does <u>not</u> always hold:

(4.4.2) Problem. Let $S = \{1, 2\}$, with $p_{11}^{(0)} = p_{22}^{(0)} = 1$. Suppose that for all $t > 0$, $p_{11}^{(t)} = p_{21}^{(t)} = 1$.
(a) Verify that these $p_{ij}^{(t)}$ are valid transition probabilities.
(b) Verify the Chapman-Kolmogorov equations for these $p_{ij}^{(t)}$.
(c) Show that the corresponding process is <u>not</u> standard.

From now on, we will <u>assume</u> that our processes are standard.

Since $p_{ii}^{(0)} = 1 > 0$, it follows that for all small enough $h > 0$, $p_{ii}^{(h)} > 0$. Then, by the Chapman-Kolmogorov inequality, we have $p_{ii}^{(nh)} > 0$ for all small enough $h > 0$ and all $n \in \mathbf{N}$. Hence, $p_{ii}^{(t)} > 0$ for all $t \geq 0$. It follows that continuous-time standard processes are <u>always aperiodic</u>.

A new concept is:

(4.4.3) Definition. Given a standard Markov process, its *generator* is $g_{ij} = \lim_{t \searrow 0} \frac{p_{ij}^{(t)} - \delta_{ij}}{t} = p_{ij}'(0)$ (where $'$ means the <u>right-handed</u> derivative), with matrix $G = (g_{ij})_{i,j \in S} = P'(0) = \lim_{t \searrow 0} \frac{P(t) - I}{t}$.

It is a fact that, for standard processes, these derivatives always exist. Interpretation: If $t > 0$ is small, then $\frac{P(t) - I}{t} \approx G$, so $P(t) \approx I + t\,G$, i.e. $p_{ij}^{(t)} \approx \delta_{ij} + t\,g_{ij}$.

As for its <u>signs</u>, clearly $g_{ii} \leq 0$, while $g_{ij} \geq 0$ for $i \neq j$.

Usually (e.g. if S is finite), we can exchange the sum and limit, so:

$$\sum_{j \in S} g_{ij} = \sum_{j \in S} \lim_{t \searrow 0} \frac{p_{ij}(t) - \delta_{ij}}{t} = \lim_{t \searrow 0} \frac{\sum_{j \in S} p_{ij}(t) - \sum_{j \in S} \delta_{ij}}{t}$$

$$= \lim_{t \searrow 0} \frac{1 - 1}{t} = 0,$$

i.e. G has row sums equal to 0.

(4.4.4) Problem. Let $\{X(t)\}_{t \geq 0}$ be a continuous-time Markov process on the state space $S = \{1, 2, 3\}$. Suppose that to first order as $t \searrow 0$, its transition probabilities are given by

$$P(t) = \begin{pmatrix} 1 - 7t & 7t & 0 \\ 0 & 1 - 3t & 3t \\ t & 2t & 1 - 3t \end{pmatrix} + o(t).$$

Compute the generator matrix G for this process.

(4.4.5) Running Example. Let $S = \{1, 2\}$, and $G = \begin{pmatrix} -3 & 3 \\ 6 & -6 \end{pmatrix}$.

Is this a valid generator? Yes!

The signs are correct (≤ 0 on the diagonal, and ≥ 0 otherwise).

And, the row sums are correct (all 0).

So, for small $t > 0$, $P^{(t)} \approx I + tG = \begin{pmatrix} 1 - 3t & 3t \\ 6t & 1 - 6t \end{pmatrix}$.

So $p_{11}^{(t)} \approx 1 - 3t$, $p_{12}^{(t)} \approx 3t$, etc.

e.g. if $t = 0.02$, then $p_{11}^{(0.02)} \doteq 1 - 3(0.02) = 0.94$,

Similarly, $p_{12}^{(0.02)} \doteq 3(0.02) = 0.06$.

And $p_{21}^{(0.02)} \doteq 6(0.02) = 0.12$, and $p_{22}^{(0.02)} \doteq 1 - 6(0.02) = 0.88$.

That is, $P^{(0.02)} \doteq \begin{pmatrix} 0.94 & 0.06 \\ 0.12 & 0.88 \end{pmatrix}$.

So, the generator G tells us $P^{(t)}$ for <u>small</u> $t > 0$. What about for larger t? Actually it tells us that, too!

(4.4.6) Continuous-Time Transitions Theorem. If a continuous-time Markov process has generator matrix G, then for any $t \geq 0$,

$$(4.4.7) \qquad P^{(t)} = \exp(tG) := I + tG + \frac{t^2 G^2}{2!} + \frac{t^3 G^3}{3!} + \cdots.$$

Proof: Recall that for $c \in \mathbf{R}$, $\lim_{n \to \infty} (1 + \frac{c}{n})^n = e^c = 1 + c + \frac{c^2}{2!} + \frac{c^3}{3!} + \cdots$

Similarly, for any matrix A, $\lim_{n \to \infty} (I + \frac{1}{n}A)^n = \exp(A)$, where we define $\exp(A) := I + A + \frac{A^2}{2!} + \frac{A^3}{3!} + \cdots$ (and I is the *identity matrix*). Hence, using the Chapman-Kolmogorov equations, for any $m \in \mathbf{N}$,

$$P^{(t)} = [P^{(t/m)}]^m = \lim_{n \to \infty} [P^{(t/n)}]^n$$

$$= \lim_{n \to \infty} [I + (t/n)G]^n = \exp(tG). \qquad \blacksquare$$

So, in principle, the generator G tells us $P^{(t)}$ for <u>all</u> $t \geq 0$. As a check, $P^{(s+t)} = \exp[(s + t)G] = \exp(sG) \exp(tG) = P^{(s)} P^{(t)}$, i.e. the Chapman-Kolmogorov equations are always satisfied. But can we actually <u>compute</u> $P^{(t)} = \exp(tG)$? Yes, sometimes!

<u>Method #1: Numerical Approximation.</u> Approximate the infinite matrix sum (4.4.7) numerically, on a computer. (Try it!)

<u>Method #2: Eigenvectors.</u> In the Running Example, if $\lambda_1 = 0$ and $\lambda_2 = -9$, and $w_1 = (2, 1)$ and $w_2 = (1, -1)$, then $w_1 G = \lambda_1 w_1 = 0$, and $w_2 G = \lambda_2 w_2 = -9 w_2$. i.e. $\{\lambda_i\}$ are <u>eigenvalues</u> of G, with <u>left-eigenvectors</u> $\{w_i\}$.

Does this help us to compute $\exp(tG)$? Yes!

If w_i is a left-eigenvector with corresponding eigenvalue λ_i, then $w_i \exp(tG) = w_i + t\lambda_i w_i + \frac{t^2\lambda_i^2}{2!} w_i + \frac{t^3\lambda_i^3}{3!} w_i + \ldots = e^{t\lambda_i} w_i$.

So, if the initial distribution is (say) $\nu = (1,0)$, then we can first find a way to write $\nu = \frac{1}{3} w_1 + \frac{1}{3} w_2$, and then compute that

$$\mu^{(t)} = \nu P^{(t)} = \nu \exp(tG) = (\frac{1}{3} w_1 + \frac{1}{3} w_2) \exp(tG)$$

$$= \frac{1}{3} e^{t\lambda_1} w_1 + \frac{1}{3} e^{t\lambda_2} w_2 = \frac{1}{3} e^{0t}(2,1) + \frac{1}{3} e^{-9t}(1,-1)$$

$$= \left(\frac{2 + e^{-9t}}{3}, \frac{1 - e^{-9t}}{3} \right).$$

So, $\mathbf{P}[X_t = 1] = p_{11}^{(t)} = \frac{2+e^{-9t}}{3}$, and $\mathbf{P}[X_t = 2] = p_{12}^{(t)} = \frac{1-e^{-9t}}{3}$.

This formula is valid for any $t \geq 0$.

As a check, $p_{11}^{(0)} = 1$, and $p_{12}^{(0)} = 0$, and $p_{11}^{(t)} + p_{12}^{(t)} = 1$.

(Or, by instead choosing $\nu = (0,1)$, we could compute $p_{21}^{(t)}$ and $p_{22}^{(t)}$.)

<u>Method #3: Differential Equation.</u> Note that

$$p'^{(t)}_{ij} = \lim_{h \searrow 0} \frac{p_{ij}^{(t+h)} - p_{ij}^{(t)}}{h} = \lim_{h \searrow 0} \frac{\left(\sum_{k \in S} p_{ik}^{(t)} p_{kj}^{(h)} \right) - p_{ij}^{(t)}}{h}$$

$$= \lim_{h \searrow 0} \frac{\left(\sum_{k \in S} p_{ik}^{(t)} [\delta_{kj} + h g_{kj}] \right) - p_{ij}^{(t)}}{h}$$

$$= \lim_{h \searrow 0} \frac{\left(p_{ij}^{(t)} + h \sum_{k \in S} p_{ik}^{(t)} g_{kj} \right) - p_{ij}^{(t)}}{h}$$

$$= \sum_{k \in S} p_{ik}^{(t)} g_{kj},$$

i.e. $P'^{(t)} = P^{(t)} G$ (which is called the *forward equations*).

As a check: $P^{(t)} = \exp(tG)$, so $P'^{(t)} = \exp(tG) G = P^{(t)} G$.

So, in the Running Example (4.4.5),

$$p'^{(t)}_{11} = p_{11}^{(t)} g_{11} + p_{12}^{(t)} g_{21} = (-3) p_{11}^{(t)} + (6) p_{12}^{(t)}$$

$$= (-3)p_{11}^{(t)} + (6)(1 - p_{11}^{(t)}) = (-9)p_{11}^{(t)} + 6 = (-9)\left(p_{11}^{(t)} - \frac{2}{3}\right).$$

But $p'^{(t)}_{11} = \frac{d}{dt}(p_{11}^{(t)}) = \frac{d}{dt}(p_{11}^{(t)} - \frac{2}{3})$. So, $\frac{d}{dt}(p_{11}^{(t)} - \frac{2}{3}) = (-9)(p_{11}^{(t)} - \frac{2}{3})$.
Hence, $p_{11}^{(t)} - \frac{2}{3} = Ke^{-9t}$, i.e. $p_{11}^{(t)} = \frac{2}{3} + Ke^{-9t}$.
But $p_{11}^{(0)} = 1$, so $K = \frac{1}{3}$, so $p_{11}^{(t)} = \frac{2}{3} + \frac{1}{3}e^{-9t} = \frac{2+e^{-9t}}{3}$.
And then $p_{12}^{(t)} = 1 - p_{11}^{(t)} = \frac{1-e^{-9t}}{3}$.
These are the same answers as before. (Phew!)

(4.4.8) Problem. Let $\{X(t)\}_{t \geq 0}$ be a continuous-time Markov process on the state space $S = \{1, 2\}$, with generator $G = \begin{pmatrix} -3 & 3 \\ 4 & -4 \end{pmatrix}$.
Compute $p_{21}^{(t)}$ for all $t > 0$, using **(a)** eigenvectors, and **(b)** a differential equation.

We next consider limiting probabilities.
In the Running Example (4.4.5), $\mu^{(t)} = (\frac{2+e^{-9t}}{3}, \frac{1-e^{-9t}}{3})$.
It follows that $\lim_{t \to \infty} \mu^{(t)} = (\frac{2}{3}, \frac{1}{3})$. So, let $\pi = (\frac{2}{3}, \frac{1}{3})$.
Then $\sum_{i \in S} \pi_i g_{i1} = \frac{2}{3}(-3) + \frac{1}{3}(6) = 0$, and $\sum_{i \in S} \pi_i g_{i2} = \frac{2}{3}(3) + \frac{1}{3}(-6) = 0$. That is, $\sum_{i \in S} \pi_i g_{ij} = 0$ for all $j \in S$, i.e. $\pi G = 0$.
Does that make sense?
Well, as in the discrete case, π should be stationary, i.e. $\sum_{i \in S} \pi_i p_{ij}^{(t)} = \pi_j$ for all $j \in S$ and all $t \geq 0$. But for small $t > 0$,

$$\sum_{i \in S} \pi_i p_{ij}^{(t)} \approx \sum_{i \in S} \pi_i [\delta_{ij} + t g_{ij}] = \pi_j + t \sum_{i \in S} \pi_i g_{ij}.$$

So, to keep π stationary, we need $\pi_j + t \sum_{i \in S} \pi_i g_{ij} = \pi_j$ for all small t, i.e. $\sum_{i \in S} \pi_i g_{ij} = 0$ for all $j \in S$. We therefore define:

(4.4.9) Definition. A probability $\{\pi_i\}$ is a *stationary* distribution for a continuous-time Markov process with generator $\{g_{ij}\}$ if $\sum_{i \in S} \pi_i g_{ij} = 0$ for all $j \in S$, i.e. if $\pi G = 0$.

In the discrete case, there was a simple reversibility condition which implied stationarity (2.2.2). In continuous time:

(4.4.10) Definition. A continuous-time Markov process with generator $\{g_{ij}\}$ is *reversible* with respect to a probability distribution $\{\pi_i\}$ if $\pi_i g_{ij} = \pi_j g_{ji}$ for all $i, j \in S$.

(4.4.11) Proposition. If a continuous-time Markov process with generator $\{g_{ij}\}$ is reversible with respect to $\{\pi_i\}$, then $\{\pi_i\}$ is a stationary distribution for the process.

Proof: We compute that $\sum_i \pi_i\, g_{ij} = \sum_i \pi_j\, g_{ji} = \pi_j \sum_i g_{ji} = \pi_j \cdot 0 = 0$, so π is stationary. ∎

So, again, reversibility implies stationarity!
In the Running Example (4.4.5), $\pi_1 g_{12} = (2/3)(3) = 2$, while $\pi_2 g_{21} = (1/3)(6) = 2$, so that example is reversible.
As before, the converse is false, i.e. it is possible to have a stationary distribution without being reversible.

(4.4.12) Problem. For a continuous-time Markov process with

$$
G = \begin{pmatrix} -3 & 3 & 0 \\ 1 & -2 & 1 \\ 0 & 4 & -4 \end{pmatrix}
$$

as its generator, find a stationary distribution π for this process, both **(a)** directly, and **(b)** using reversibility. **[sol]**

Now, even if we know that $\{\pi_i\}$ is a stationary distribution, is convergence to $\{\pi_i\}$ guaranteed?
In the discrete case, we had a theorem (2.4.1) which guaranteed convergence assuming irreducibility and aperiodicity.
In the continuous-time case, we have a similar result.
This time, we don't even need to assume aperiodicity, since standard processes are always aperiodic as discussed above.

(4.4.13) Continuous-Time Markov Convergence. If a continuous-time standard Markov process is irreducible, and has a stationary distribution π, then $\lim_{t\to\infty} p_{ij}^{(t)} = \pi_j$ for all $i, j \in S$.

Proof: Choose any $h > 0$, and let $P^{(h)} = \{p_{ij}^{(h)}\}$ be the matrix of transition probabilities in time h.
Then $P^{(h)}$ are transition probabilities for a discrete-time, irreducible, aperiodic Markov chain with stationary distribution π.
Hence, by the discrete-time Markov Chain Convergence Theorem (2.4.1), $\lim_{n\to\infty} p_{ij}^{(hn)} = \pi_j$ for any $h > 0$.

The result then follows from this *(formally, since it can be shown[12] that $\sum_j |p_{ij}^{(t)} - \pi_j|$ is a non-increasing function of t).* ∎

(4.4.14) Problem. Let $\{X(t)\}_{t\geq 0}$ be a continuous-time Markov process on the state space $S = \{1, 2\}$, with generator

$$G = \begin{pmatrix} -7 & 7 \\ 3 & -3 \end{pmatrix}.$$

(a) Compute $p_{22}^{(t)}$ for all $t \geq 0$. [Hint: You may use without proof (if you wish) that the eigenvalues of G are $\lambda_1 = 0$ and $\lambda_2 = -10$, with corresponding left-eigenvectors $v_1 = (3, 7)$ and $v_2 = (1, -1)$, and furthermore $(0, 1) = \frac{1}{10}(v_1 - 3v_2)$.]
(b) Compute $\lim_{t\to\infty} p_{22}^{(t)}$.

(4.4.15) Problem. Let $\{X(t)\}_{t\geq 0}$ be a continuous-time Markov process on the state space $S = \{1, 2\}$, with $\nu_1 = 1/4$ and $\nu_2 = 3/4$, and with generator given by

$$G = \begin{pmatrix} -2 & 2 \\ 3 & -3 \end{pmatrix}.$$

(a) Compute $\mathbf{P}[X(t) = 2]$ to first order as $t \searrow 0$.
(b) Compute $\mathbf{P}[X(t) = 2]$ for all $t > 0$.   [Hint: You might be interested in the vectors $w_1 = (1, -1)$ and $w_2 = (3, 2)$.]
(c) Compute the limiting probability $\lim_{t\to\infty} \mathbf{P}[X(t) = 2]$.

(4.4.16) Problem.   Let $\{X(t)\}_{t\geq 0}$ be a continuous-time Markov process on the state space $S = \{1, 2, 3\}$. Suppose that to first order as $t \searrow 0$, its transition probabilities are given by

$$P^{(t)} = \begin{pmatrix} 1 - 7t & 7t & 0 \\ 0 & 1 - 4t & 4t \\ t & 2t & 1 - 3t \end{pmatrix} + o(t).$$

(a) Compute the generator matrix G for this process.

---

[12] See e.g. Proposition 3(c) of Roberts and Rosenthal (2004).

(b) Find a stationary distribution $\{\pi_i\}$ for this process.

(c) Determine whether or not $\lim_{t\to\infty} p_{ij}^{(t)} = \pi_j$ for all $i, j \in S$.

(4.4.17) Problem. Let $G_1 = \begin{pmatrix} 2 & 5 \\ -2 & -5 \end{pmatrix}$, $G_2 = \begin{pmatrix} -5 & 5 \\ 2 & -2 \end{pmatrix}$,
$G_3 = \begin{pmatrix} 2 & -2 \\ 5 & -5 \end{pmatrix}$, $G_4 = \begin{pmatrix} 2 & -5 \\ -2 & 5 \end{pmatrix}$, $G_5 = \begin{pmatrix} 2 & -2 \\ -5 & 5 \end{pmatrix}$, and $G_6 = \begin{pmatrix} -5 & 2 \\ 5 & -2 \end{pmatrix}$.

(a) Specify (with explanation) which <u>one</u> of G_1, G_2, G_3, G_4, G_5, and G_6 is a valid generator for a Markov process $\{X(t)\}$ on $S = \{1, 2\}$.

(b) For the generator selected in part (a), compute $\mathbf{P}_2[X(t) = 2]$ for all $t > 0$. [Hint: Vectors $w_1 = (-1, 1)$ and $w_2 = (2, 5)$ might help.]

(c) For the generator from part (a), compute $\lim_{t\to\infty} \mathbf{P}_2[X(t) = 2]$.

Connections to Discrete-Time Markov Chains:

We next consider connections between discrete-time Markov chains and continuous-time Markov processes.

Let $\{\hat{X}_n\}_{n=0}^{\infty}$ be a discrete-time chain with transitions $\{\hat{p}_{ij}\}_{i,j\in S}$. And, let $\{N(t)\}_{t\geq 0}$ be a Poisson process with intensity $\lambda > 0$. Then, let $X_t = \hat{X}_{N(t)}$. So, $\{X_t\}$ is just like $\{\hat{X}_n\}$, except that it jumps at Poisson process event times, not at integer times. Here $\{X_t\}$ is the rate λ *exponential holding times* version of $\{\hat{X}_n\}_{n=0}^{\infty}$. In particular, $\{X_t\}$ is a continuous-time Markov process. That is, we have "created" a continuous-time Markov process from a discrete-time Markov chain. We can then compute the <u>generator</u> of this Markov process $\{X_t\}$:

(4.4.18) Proposition. The rate λ exponential holding times version of a discrete-time Markov chain with transitions $\{\hat{p}_{ij}\}_{i,j\in S}$ has generator $G = \lambda(\hat{P} - I)$, i.e. $g_{ii} = \lambda(\hat{p}_{ii} - 1)$, and for $i \neq j$, $g_{ij} = \lambda \hat{p}_{ij}$.

Proof: By conditioning (A.6.5), $p_{ij}^{(t)} = \mathbf{P}_i[\hat{X}_{N(t)} = j] = \sum_{n=0}^{\infty} \mathbf{P}_i[N(t) = n, \hat{X}_n = j] = \sum_{n=0}^{\infty} \mathbf{P}[N(t) = n] \hat{p}_{ij}^{(n)}$. So, as $t \searrow 0$,

$$p_{ij}^{(t)} = \mathbf{P}[N(t) = 0] \hat{p}_{ij}^{(0)} + \mathbf{P}[N(t) = 1] \hat{p}_{ij}^{(1)} + \mathbf{P}[N(t) = 2] \hat{p}_{ij}^{(2)} + \dots$$

$$= [e^{-\lambda t}(\lambda t)^0/0!] \delta_{ij} + [e^{-\lambda t}(\lambda t)^1/1!] \hat{p}_{ij} + O(t^2)$$

$$= (1 - \lambda t)\,\delta_{ij} + (1 - \lambda t)(\lambda t)\,\hat{p}_{ij} + O(t^2)$$
$$= \delta_{ij} + t[\lambda(\hat{p}_{ij} - \delta_{ij})] + O(t^2)\,.$$

But as $t \searrow 0$, $p_{ij}^{(t)} = \delta_{ij} + tg_{ij} + O(t^2)$.

Hence, we must have $tg_{ij} = t[\lambda(\hat{p}_{ij} - \delta_{ij})]$,

So, $g_{ij} = \lambda\,(\hat{p}_{ij} - \delta_{ij})$, which gives the result. ∎

As a check: $g_{ii} \leq 0$, and for $i \neq j$, $g_{ij} \geq 0$, and the row sums are $\sum_{j \in S} g_{ij} = g_{ii} + \sum_{j \neq i} g_{ij} = \lambda(\hat{p}_{ii} - 1) + \sum_{j \neq i}(\lambda \hat{p}_{ij}) = -\lambda + \sum_{j \in S}(\lambda \hat{p}_{ij}) = -\lambda + \lambda \sum_{j \in S} \hat{p}_{ij} = -\lambda + \lambda(1) = 0$, as they must be.

One special case is when $\hat{X}_0 = 0$, and $\hat{p}_{i,i+1} = 1$ for all i, i.e.

$$\hat{P} \;=\; \begin{pmatrix} 0 & 1 & 0 & 0 & \cdots \\ 0 & 0 & 1 & 0 & \cdots \\ 0 & 0 & 0 & 1 & \cdots \end{pmatrix}.$$

That is, $\hat{X}_n = n$ for all $n \geq 0$.

This is a rather boring discrete-time Markov chain!

However, it follows that $X_t = \hat{X}_{N(t)} = N(t)$.

That is, $\{X_t\}$ is itself the Poisson process $\{N(t)\}$.

Hence, Proposition (4.4.18) tells us:

(4.4.19) Corollary. A Poisson process $\{N(t)\}_{t \geq 0}$ with rate $\lambda > 0$ has generator given by:

$$G \;=\; \lambda(\hat{P} - I) \;=\; \begin{pmatrix} -\lambda & \lambda & 0 & 0 & \cdots \\ 0 & -\lambda & \lambda & 0 & \cdots \\ 0 & 0 & -\lambda & \lambda & \cdots \end{pmatrix}.$$

We can also make a connection the other way:

Suppose $\{X_t\}$ is a continuous-time process.

Then let $\{\hat{X}_n\}$ be the corresponding *jump chain*, i.e. the discrete-time Markov chain consisting of each new state that $\{X_t\}$ visits.

That is, to define $\{\hat{X}_n\}$, we ignore how much <u>time</u> $\{X_t\}$ spends at each state, and just list all the new states $\{X_t\}$ visits, in order.

(So, $\{\hat{X}_n\}$ might return to some state again later, but it will never be at the same state twice in a row.)

Then we have:

(4.4.20) Proposition. If a continuous-time Markov process $\{X_t\}$ has generator G, then its corresponding jump chain $\{\hat{X}_n\}$ has transition probabilities $\hat{p}_{ii} = 0$, and for $j \neq i$,

$$\hat{p}_{ij} = \frac{g_{ij}}{\sum_{k \neq i} g_{ik}}.$$

Proof: Since the jump chain never stays at the same state, $\hat{p}_{ii} = 0$. For $j \neq i$, $\hat{p}_{ij}$ is the probability that $\{X_t\}$ will jump to j when it first leaves i. How can we compute this?

Well, multiple jumps have probabilities of smaller order as $t \searrow 0$, so as $t \searrow 0$ we can assume there is at most one jump. Then

$$\hat{p}_{ij} = \lim_{t \searrow 0} \mathbf{P}_i(X_t = j \mid X_t \neq i) = \lim_{t \searrow 0} \frac{\mathbf{P}_i(X_t = j)}{\sum_{k \neq i} \mathbf{P}_i(X_t = k)}.$$

But as $t \searrow 0$, $\mathbf{P}_i(X_t = j) = p_{ij}^{(t)} \approx t\, g_{ij}$. Hence,

$$\hat{p}_{ij} = \lim_{t \searrow 0} \frac{t\, g_{ij}}{\sum_{k \neq i} t\, g_{ik}} = \frac{g_{ij}}{\sum_{k \neq i} g_{ik}}. \qquad \blacksquare$$

4.5. Application – Queueing Theory

We now consider modeling a <u>queue</u>, i.e. a line-up of customers. Specifically, we assume that customers arrive at certain times, and are then served one at a time (i.e., there is only one *server*).

Let T_n be the time of <u>arrival</u> of the nth customer.

Let $Y_n = T_n - T_{n-1}$ be the *interarrival times* between the $(n-1)^{\text{st}}$ and n^{th} customers (with the convention that $T_0 = 0$).

Let S_n be the time it takes to serve the nth customer.

We assume the $\{Y_n\}$ and $\{S_n\}$ are all independent.

Let $Q(t)$ be the *queue size*, i.e. the number of customers who are either waiting in the queue or being served at each time $t \geq 0$.

What happens to $Q(t)$ as $t \to \infty$?

Does $Q(t)$ converge? remain bounded? increase to infinity? etc.

(Such questions are important since queue processes are used to model communication networks, internet routing, workflows, transportation systems, etc.; $Q(t) \to \infty$ indicates instability while bounded $Q(t)$ indicates stability.)

Consider first an *M/M/1 queue*, with $T_n - T_{n-1} \sim$ Exponential(λ), and $S_n \sim$ Exponential(μ), for some $\lambda, \mu > 0$.
Then $\{T_n\}$ are arrival times of a Poisson process with intensity λ.
But the departure times are more complicated, since customers can't be served until they arrive (and the server is available).
Still, by the memoryless property (A.6.9), $\{Q(t)\}$ is a Markov process. (Indeed, both of the "M" in "M/M/1" stand for "Markov".)
So, we can find its *generator*.

(4.5.1) Proposition.   An M/M/1 queue with interarrival times $T_n - T_{n-1} \sim$ Exponential(λ) and service times $S_n \sim$ Exponential(μ) has generator matrix given by:

$$
G = \begin{pmatrix}
-\lambda & \lambda & 0 & 0 & 0 & \cdots \\
\mu & -\lambda - \mu & \lambda & 0 & 0 & \cdots \\
0 & \mu & -\lambda - \mu & \lambda & 0 & \cdots \\
0 & 0 & \ddots & \ddots & \ddots
\end{pmatrix}.
$$

Proof: By definition,

$$
g_{n,m} = \lim_{t \searrow 0} \frac{\mathbf{P}[Q(t) = m \mid Q(0) = n]}{t}
$$

So, we need to compute $\mathbf{P}[Q(t) = m \mid Q(0) = n]$ to first order in t, as $t \searrow 0$.

Well, if $|n - m| \geq 2$, then to change $Q(t)$ from n to m requires at least <u>two</u> arrivals or at least <u>two</u> departures, each of which has probability $o(t)$ by Proposition (4.3.9)(b), so $g_{n,m} = \lim_{t \searrow 0} \frac{o(t)}{t} = 0$.

Next, if $m = n + 1$, then to change $Q(t)$ from n to $n+1$ requires <u>either</u> at least two arrivals plus at least one departure (which again has probability $o(t)$ and can be ignored), <u>or</u> exactly one arrival and zero departures.
If $n \geq 1$, then using (A.4.9), one arrival and zero departures has probability equal to

$(e^{-\lambda t}\lambda t)(e^{-\mu t}(1)) = [1 - \lambda t + O(t^2)]\lambda t[1 - \mu t + O(t^2)] = \lambda t + O(t^2)$,

so $g_{n,n+1} = \lim_{t \searrow 0} \frac{\lambda t + O(t^2)}{t} = \lambda$.

If $n = 0$, then this probability is somewhat more complicated, since an empty queue never has departures. However, it is still somewhere between $(e^{-\lambda t}\lambda t)(e^{-\mu t}(1))$ (if the queue were always full) and $(e^{-\lambda t}\lambda t)(1)$ (if the queue were always empty). So, it is still $\lambda t + O(t^2)$, so still $g_{01} = \lambda$.

If instead $m = n - 1$ where $n \geq 1$, then to change $Q(t)$ from n to $n - 1$ requires <u>either</u> at least two departures and one arrival (with probability $o(t)$ which can be ignored), <u>or</u> exactly one departure and zero arrivals.

If $n \geq 2$, then using (A.4.9), one departure and zero arrivals has probability equal to

$(e^{-\mu t}\mu t)(e^{-\lambda t}(1)) = [1 - \mu t + O(t^2)]\mu t[1 - \lambda t + O(t^2)] = \mu t + O(t^2)$,

so $g_{n,n-1} = \lim_{t \searrow 0} \frac{\mu t + O(t^2)}{t} = \mu$.

If $n = 1$, then the probability is somewhat more complicated, since an empty queue never has departures. However, the probability is still somewhere between the probability $(e^{-\mu t}\mu t)(e^{-\lambda t}(1)) = \mu t + O(t^2)$ of <u>exactly</u> one attempted departure, and $(1 - e^{-\mu t}(1))(e^{-\lambda t}(1)) = (1 - [1 - \mu t + O(t^2)])[1 - \lambda t + O(t^2)] = (\mu t + O(t^2))[1 - \lambda t + O(t^2)] = \mu t + O(t^2)$ of at <u>least</u> one attempted departure. So, it is still $\mu t + O(t^2)$, so still $g_{10} = \mu$.

It remains to compute g_{nn} for all n.

However, we must have row sums $\sum_{m=0}^{\infty} g_{n,m} = 0$.

So, we must have $g_{00} = -\lambda$, and $g_{nn} = -\lambda - \mu$ for $n \geq 1$. ∎

(4.5.2) Problem. Consider an $M/M/1$ single-server queue˙with $\lambda = 6$, and $\mu = 8$, and set $p_{ij}(h) = \mathbf{P}[Q(h) = j \,|\, Q(0) = i]$.
(a) Find a value $r \geq 0$ with $0 < \lim_{h \searrow 0} [p_{57}(h) / h^r] < \infty$. [sol]
(b) For this r, compute (with explanation) $\lim_{h \searrow 0} [p_{57}(h) / h^r]$. [sol]

So, we have found the M/M/1 queue's generator matrix G. We can now use that to compute its stationary distribution:

(4.5.3) Proposition. If $\lambda < \mu$, an M/M/1 queue with interarrival times $T_n - T_{n-1} \sim$ Exponential(λ) and service times $S_n \sim$

Exponential(μ) has a geometric (A.3.3) stationary distribution:

$$\pi_i = \left(\frac{\lambda}{\mu}\right)^i \left(1 - \frac{\lambda}{\mu}\right), \qquad i = 0, 1, 2, 3, \ldots .$$

Proof: We need $\pi G = 0$, i.e. $\sum_{i \in S} \pi_i g_{ij} = 0$ for all $j \in S$.
Setting $j = 0$ gives: $\pi_0(-\lambda) + \pi_1(\mu) = 0$, so $\pi_1 = \left(\frac{\lambda}{\mu}\right)\pi_0$.
Setting $j = 1$ gives: $\pi_0(\lambda) + \pi_1(-\lambda - \mu) + \pi_2(\mu) = 0$.
So $\pi_2 = \left(\frac{\lambda}{-\mu}\right)\pi_0 + \left(\frac{-\lambda-\mu}{-\mu}\right)\pi_1 = \left(-\frac{\lambda}{\mu}\right)\pi_0 + \left(1 + \frac{\lambda}{\mu}\right)\left(\frac{\lambda}{\mu}\right)\pi_0 = \left(\frac{\lambda}{\mu}\right)^2\pi_0$.
Continuing by induction gives $\pi_i = \left(\frac{\lambda}{\mu}\right)^i \pi_0$, for $i = 0, 1, 2, \ldots$
Then since $\lambda < \mu$ and $\sum_i \pi_i = 1$, we have by (A.4.1) that

$$\pi_0 = \frac{1}{\sum_{i=0}^{\infty}(\frac{\lambda}{\mu})^i} = \frac{1}{\frac{1}{1-\frac{\lambda}{\mu}}} = 1 - \frac{\lambda}{\mu}.$$

So, $\pi_i = \left(\frac{\lambda}{\mu}\right)^i \pi_0 = \left(\frac{\lambda}{\mu}\right)^i\left(1 - \frac{\lambda}{\mu}\right)$, as claimed. ∎

(4.5.4) Problem. Show that an M/M/1 queue with $\lambda < \mu$ is reversible with respect to the π in Proposition (4.5.3).

Proposition (4.5.3) says that if $\lambda < \mu$, i.e. $\frac{1}{\mu} < \frac{1}{\lambda}$, i.e. $\mathbf{E}(S_n) < \mathbf{E}(Y_n)$, then the M/M/1 queue has a stationary distribution.
And, the queue is irreducible (since it can always add or subtract 1).
So, by the Continuous-Time Markov Convergence Theorem (4.4.13):

(4.5.5) Proposition. If $\lambda < \mu$, the queue size $Q(t)$ of an M/M/1 queue with interarrival times $T_n - T_{n-1} \sim$ Exponential(λ) and service times $S_n \sim$ Exponential(μ) has limiting probabilities

$$\lim_{t \to \infty} \mathbf{P}[Q(t) = i] = \pi_i = \left(\frac{\lambda}{\mu}\right)^i \left(1 - \frac{\lambda}{\mu}\right).$$

(4.5.6) Problem. Consider an $M/M/1$ queue with $\lambda = 3$, and $\mu = 5$. Compute $\lim_{t \to \infty} \mathbf{P}[Q(t) = 2]$.

It follows from Proposition (4.5.5) that

$$\lim_{t \to \infty} \mathbf{P}[Q(t) \geq K] = \sum_{i=K}^{\infty} \left(\frac{\lambda}{\mu}\right)^i \left(1 - \frac{\lambda}{\mu}\right) = \left(\frac{\lambda}{\mu}\right)^K.$$

Hence, $\lim_{K\to\infty} \lim_{t\to\infty} \mathbf{P}[Q(t) \geq K] = \lim_{K\to\infty} (\frac{\lambda}{\mu})^K = 0$.
That is, the collection $\{Q(t)\}_{t\geq 0}$ is *bounded in probability.*

However, if instead $\lambda > \mu$, i.e. $\mathbf{E}(S_n) > \mathbf{E}(Y_n)$, then there is no
stationary distribution. Then what happens to $Q(t)$ as $t \to \infty$?
Well, in this case, the jump chain (4.4.20) for $Q(t)$ is essentially s.r.w.
with $p = \frac{\lambda}{\lambda+\mu} > 1/2$ (except that it can never go negative, but that
only makes it <u>larger</u>).
So, as in Proposition (1.6.17), the jump chain converges to positive
infinity w.p. 1 (and thus also in probability).
So, $Q(t)$ also converges to positive infinity w.p. 1 and in probability.
Therefore, $\lim_{t\to\infty} \mathbf{P}[Q(t) \geq K] = 1$ for all K.
Hence, $\lim_{K\to\infty} \lim_{t\to\infty} \mathbf{P}[Q(t) \geq K] = 1$.
That is, $\{Q(t)\}_{t\geq 0}$ is <u>not</u> bounded in probability.

If $\lambda = \mu$, i.e. $\mathbf{E}(S_n) = \mathbf{E}(Y_n)$, the jump chain (4.4.20) for $Q(t)$
corresponds precisely to the absolute value of s.s.r.w.
So, as in (2.4.10), the jump chain, and hence also $Q(t)$ itself, con-
verges to positive infinity in probability (though not w.p. 1).
Hence, again $\lim_{K\to\infty} \lim_{t\to\infty} \mathbf{P}[Q(t) \geq K] = 1$.
So, again, $\{Q(t)\}_{t\geq 0}$ is <u>not</u> bounded in probability.

In fact, these last observations are true much more generally!
Consider instead a (single-server) *G/G/1 queue:*
This means that we do <u>not</u> assume Exponential distributions, just
that the interarrival times $\{Y_n\} = \{T_n - T_{n-1}\}$ are i.i.d., and also
the service times $\{S_n\}$ are i.i.d., and they are all independent.
This is much more general than the M/M/1 case.
(Indeed, both of the "G" in "G/G/1" stand for "General".)
Here $Q(t)$ is <u>not</u> Markovian, so we cannot compute generators, nor
stationary distributions, nor limiting probabilities.
However, we still have the following:

(4.5.7) General Queue Theorem. For a G/G/1 queue:
(a) If $\mathbf{E}(S_n) > \mathbf{E}(Y_n)$, then $Q(t) \to \infty$ w.p. 1, and hence also $Q(t) \to$
∞ in probability, So, $\{Q(t)\}_{t\geq 0}$ is <u>not</u> bounded in probability. (This
corresponds to the M/M/1 case $\lambda > \mu$.)
(b) If $\mathbf{E}(S_n) < \mathbf{E}(Y_n)$, then $\{Q(t)\}_{t\geq 0}$ is bounded in probability, i.e.
$\lim_{K\to\infty} \limsup_{t\to\infty} \mathbf{P}(Q(t) \geq K) = 0$. (This corresponds to the

M/M/1 case $\lambda < \mu$.)
(c) If $\mathbf{E}(S_n) = \mathbf{E}(Y_n)$, and the Y_n and S_n are not both constant (i.e. $\mathbf{Var}(Y_n) + \mathbf{Var}(S_n) > 0$), then $Q(t) \to \infty$ in probability but not w.p. 1. (This corresponds to the borderline M/M/1 case $\lambda = \mu$.)

(4.5.8) Problem. Consider a G/G/1 queue with i.i.d. interarrival times $Y_n \sim$ Exponential(5), and i.i.d. service times $S_n \sim$ Uniform$[0, C]$ for some $C > 0$. Specify (with explanation) precisely which values of C do and do not make the queue sizes $\{Q(t)\}_{t \geq 0}$ be bounded in probability. **[sol]**

We do not prove Theorem (4.5.7) here[13]. However, we next prove (most of) a similar result about waiting times.

Waiting Times: Let W_n be the amount of time the nth customer has to <u>wait</u> before being served. Then, does $W_n \to \infty$?

Well, for an M/M/1 queue with $\lambda < \mu$, by (4.5.5) the number of people in the queue converges to Geometric(λ/μ).
Hence, for large n, $\mathbf{P}(W_n > K)$ converges to the probability that a sum of Geometric(λ/μ) i.i.d. variables will exceed K.
More precisely, since a sum of i.i.d. Exponential random variables is Gamma, for an M/M/1 queue with $\lambda < \mu$ we have

$$\lim_{n \to \infty} \mathbf{P}(W_n > K) = \sum_{i=0}^{\infty} \left(\frac{\lambda}{\mu}\right)^i \left(1 - \frac{\lambda}{\mu}\right) \Gamma(i, \mu; K) < \infty,$$

where $\Gamma(i, \mu; K) = \mathbf{P}[$a $\Gamma(i, \mu)$ random variable is $> K]$.
In particular, $\lim_{K \to \infty} \lim_{n \to \infty} \mathbf{P}(W_n > K) = 0$, i.e. the waiting times $\{W_n\}$ are *bounded in probability*.

By contrast, if $\lambda \geq \mu$, then $Q(t) \to \infty$.
Does this mean that also $W_n \to \infty$?

---

[13]See e.g. Section 3.2.1 of the book "Stochastic Models of Manufacturing Systems" by J.A. Buzacott and J.G. Shanthikumar, Prentice-Hall, 1993. Their proof proceeds by first showing that the queue will empty infinitely often with probability 1. I am very grateful to Opher Baron for pointing me to this reference and for helpful discussions.

Yes! And a similar result holds for the more general G/G/1 case. Indeed, we have[14] that:

(4.5.9) General Queue Waiting Time Theorem. For a general G/G/1 single-server queue,
(a) If $\mathbf{E}(Y_n) < \mathbf{E}(S_n)$, then $W_n \to \infty$ w.p. 1.
(b) If $\mathbf{E}(Y_n) > \mathbf{E}(S_n)$, then $\{W_n\}$ is bounded in probability.
(c) If $\mathbf{E}(Y_n) = \mathbf{E}(S_n)$, and $\mathbf{Var}(S_{n-1}) + \mathbf{Var}(Y_n) > 0$, then $W_n \to \infty$ in probability (but <u>not</u> w.p. 1).

To prove (4.5.9) requires some technical material about waiting times:

(4.5.10) Lindley's Equation. *For $n \geq 1$,*

$$W_n = \max(0, \ W_{n-1} + S_{n-1} - Y_n).$$

Proof:
The $(n-1)^{\text{st}}$ customer is in the system for time $W_{n-1} + S_{n-1}$.
But the nth customer arrives Y_n <u>after</u> the $(n-1)^{\text{st}}$ customer.
If $W_{n-1} + S_{n-1} \leq Y_n$, then the nth customer doesn't have to wait at all, so $W_n = 0$, consistent with the claim.
Or, if $W_{n-1} + S_{n-1} \geq Y_n$, then $W_n = $ [time the $(n-1)^{\text{st}}$ customer is in the system] $-$ [amount of that time that the nth customer was <u>not</u> present for] $= (W_{n-1} + S_{n-1}) - Y_n$. ∎

(4.5.11) Lindley's Corollary. $\displaystyle W_n = \max_{0 \leq k \leq n} \sum_{i=k+1}^{n} (S_{i-1} - Y_i)$
(where if $k = n$ then the sum equals zero).

Proof #1: *Write it out: $W_0 = 0$, $W_1 = \max(0, S_0 - Y_1)$,*
$W_2 = \max(0, W_1 + S_1 - Y_2) = \max(0, \max(0, S_0 - Y_1) + S_1 - Y_2) = \max(0, \ S_1 - Y_2, \ S_0 - Y_1 + S_1 - Y_2)$,
$W_3 = \max(0, W_2 + S_2 - Y_3) = \max(0, \max(0, \ S_1 - Y_2, \ S_0 - Y_1 + S_1 - Y_2) + S_2 - Y_3) = \max(0, S_2 - Y_3, \ S_1 - Y_2 + S_2 - Y_3, \ S_0 - Y_1 + S_1 - Y_2 + S_2 - Y_3)$, etc., each corresponding to the claimed formula.

Proof #2: *Induction on n. When $n = 0$, both sides are zero. If n increases to $n + 1$, then by Lindley's Equation, each possible*

---

[14]This result was first proved by D.V. Lindley, "The theory of queues with a single server", Math. Proc. Cam. Phil. Soc. **48(2)** (1952), 277–289. It was then generalised to the case of $s \geq 1$ servers, with $\mathbf{E}(Y_n)$ replaced by $s\,\mathbf{E}(Y_n)$, by J. Kiefer and J. Wolfowitz, "On the Theory of Queues With Many Servers", Trans. Amer. Math. Soc. **78(1)**, 1–18.

value of the "max" gets $S_n - Y_{n+1}$ added to it. And the "max with zero" is covered by allowing for the possibility $k = n + 1$. ∎

Proof of Theorem (4.5.9):

For part (a), by Lindley's Equation, $W_{n+1} \geq W_n + S_n - Y_{n+1}$. Here the sequence $\{S_n - Y_{n+1}\}$ is i.i.d., with mean > 0. So, by the Law of Large Numbers (A.7.2), $\lim_{n \to \infty} \frac{W_n}{n} \geq \mathbf{E}(S_n - Y_{n+1}) > 0$, w.p. 1. It follows that $\lim_{n \to \infty} W_n \geq \infty$, w.p. 1.

For part (b), by Lindley's Corollary,
$$\mathbf{P}(W_n > K) = \mathbf{P}\left(\max_{0 \leq k \leq n} \sum_{i=k+1}^{n}(S_{i-1} - Y_i) > K \right).$$
But $\{S_{i-1} - Y_i\}$ are i.i.d., so this is equivalent to
$$\mathbf{P}(W_n > K) = \mathbf{P}\left(\max_{0 \leq k \leq n} \sum_{i=1}^{n-k}(S_{i-1} - Y_i) > K \right).$$
This is the probability that i.i.d. partial sums with negative mean (since $\mathbf{E}(S_{i-1} - Y_i) < 0$) will ever be larger than K, i.e. it is the probability that the <u>maximum</u> of a sequence of i.i.d. partial sums with negative mean will be larger than K.
But by the Law of Large Numbers (A.7.2), i.i.d. partial sums with negative mean will eventually become and remain <u>negative</u>, w.p. 1. So, w.p. 1, only a <u>finite</u> number of the partial sums have positive values, so their maximum value is <u>finite</u>.
So, as $K \to \infty$, the probability that the maximum value will be $> K$ must converge to zero.
So, for any $\epsilon > 0$, there is $K < \infty$ such that the probability that its maximum value will be $> K$ is $< \epsilon$.

Part (c) is analogous to (2.4.10) but more technical; see e.g. Grimmett & Stirzaker (1992), Theorem 11.5(4)(c), pp. 432–435. ∎

(4.5.12) Problem. Consider a single-server queue. Suppose the 3rd customer waits for 10 minutes, and then takes 3 minutes to be served. The 4th customer arrives 2 minutes after the 3rd customer. How long does the 4th customer have to wait before being served? [Hint: Don't forget Lindley's Equation.]

4.6. Application – Renewal Theory

We now consider processes which sometimes "renew" themselves. We write the *renewal times* as $\{T_n\}$.

We take $T_0 = 0$, and $T_n = Y_1 + Y_2 + \ldots + Y_n$, where $\{Y_n\}_{n=1}^{\infty}$ are independent interarrival times, with at least $\{Y_n\}_{n=2}^{\infty}$ (i.e., all but Y_1) identical distributed (i.e., i.i.d.).

Assume that the $\{Y_n\}$ are finite, i.e. $\mathbf{P}(Y_n < \infty) = 1$ for all n.

Let $N(t) = \max\{n \geq 0 : T_n \leq t\} = \#\{n \geq 1 : T_n \leq t\}$ be the number of renewal times up to time t.

Then $\{N(t)\}_{t \geq 0}$ is a *renewal process* on the state space $S = \{0, 1, 2, \ldots\}$.

If $\{Y_n\}_{n=1}^{\infty}$ are all i.i.d., then the process is *zero-delayed*.

If $\{Y_n\}$ are i.i.d. $\sim \mathrm{Exp}(\lambda)$, then $\{N(t)\}$ is Markovian (by the memoryless property), and in fact $\{N(t)\}$ is a Poisson process.

But for other distributions, usually $\{N(t)\}$ is not Markovian.

(In fact, Poisson processes are the only Markovian renewal processes, cf. Grimmett & Stirzaker (1992), Theorem 8.3(5), p. 339.)

(4.6.1) Example. Suppose we have a light bulb, which we replace immediately whenever it burns out.

Let T_n be the nth time we replace it (with $T_0 = 0$.)

Then $Y_n = T_n - T_{n-1}$ is the lifetime of the nth bulb, for $n \geq 2$.

If the bulbs are identical, then $\{Y_n\}_{n=2}^{\infty}$ are i.i.d.

Let $N(t)$ be the number of bulbs replaced by time t.

Then $\{N(t)\}$ is a renewal process.

If we started with a fresh bulb, then $\{Y_n\}_{n=1}^{\infty}$ are all i.i.d., so $\{N(t)\}$ is zero-delayed, otherwise it's probably not.

Similar examples arise for repeatedly fixing a machine, etc.

(4.6.2) Example. Let $\{Q(t)\}$ be a single-server queue.

Let $T_0 = 0$, and let T_n be the nth time the queue empties (i.e., the nth time s such that $Q(s) = 0$ but $\lim_{t \nearrow s} Q(t) > 0$).

Let $Y_n = T_n - T_{n-1}$ be the time between queue emptyings.

Let $N(t) = \#\{n \geq 1 : T_n \leq t\}$ be the number of times the queue has emptied up to time t.

Then $\{N(t)\}$ is a renewal process.

It is probably not zero-delayed, unless we start right when the queue empties, or else start at some time when the queue is empty and also have Exponentially-distributed interarrival times.

(4.6.3) Example. Let $\{X_n\}_{n=0}^{\infty}$ be an irreducible recurrent discrete-time Markov chain on a discrete state space S.

Let $i, j \in S$, and assume $X_0 = j$.
Let $T_0 = 0$, and for $n \geq 1$, let $T_n = \min\{k > T_{n-1} : X_k = i\}$ be the time of the nth visit to the state i.
Let $Y_n = T_n - T_{n-1}$. Then for $y = 1, 2, \ldots$, we have

$$\mathbf{P}(Y_1 = y) = \mathbf{P}_j(X_y = i, \text{ but } X_m \neq i \text{ for } 1 \leq m \leq y - 1),$$

and for $n \geq 2$,

$$\mathbf{P}(Y_n = y) = \mathbf{P}_i(X_y = i, \text{ but } X_m \neq i \text{ for } 1 \leq m \leq y - 1).$$

Let $N(t) = \#\{n \geq 1 : T_n \leq t\}$ be the number of visits to i by time t. Then $\{N(t)\}$ is a renewal process.
If $j = i$, the process is zero-delayed, otherwise it's probably not.

So, there are lots of examples of renewal processes.
We often want to know about $N(t)$ or $N(t)/t$ for large t.
Some properties depend on the *mean interarrival time* $\mu := \mathbf{E}(Y_2) = \mathbf{E}(Y_3) = \ldots$ One such result is the:

(4.6.4) Elementary Renewal Theorem. For a renewal process as above, if $\mu < \infty$, then:

(a) $\lim_{t \to \infty} \frac{N(t)}{t} = 1/\mu$ w.p. 1, and

(b) $\lim_{t \to \infty} \frac{\mathbf{E}[N(t)]}{t} = 1/\mu$.

Proof: For (a), by the Law of Large Numbers (A.7.2), w.p. 1,
$\lim_{n \to \infty} \frac{1}{n} T_n = \lim_{n \to \infty} \frac{1}{n}(Y_1 + Y_2 + \ldots + Y_n) = \mu$.
But since $Y_n < \infty$ w.p. 1 for all n, $\lim_{t \to \infty} N(t) = \infty$ w.p. 1.
Hence, $\lim_{t \to \infty} \frac{1}{N(t)} T_{N(t)} = \lim_{n \to \infty} \frac{1}{n} T_n = \mu$ w.p. 1.
But $N(t) = \max\{n \geq 0 : T_n \leq t\}$, so $T_{N(t)} \leq t < T_{N(t)+1}$.
We then have the string of relationships:

$$\frac{T_{N(t)}}{N(t)} \leq \frac{t}{N(t)} < \frac{T_{N(t)+1}}{N(t)} = \frac{T_{N(t)+1}}{N(t)+1} \frac{N(t)+1}{N(t)}.$$

As $t \to \infty$, w.p. 1, $\frac{T_{N(t)}}{N(t)} \to \mu$, and $\frac{T_{N(t)+1}}{N(t)+1} \to \mu$, and $\frac{N(t)+1}{N(t)} \to 1$.
So, $\frac{t}{N(t)} \to \mu$ w.p. 1 by the Sandwich Theorem (A.4.8).
Taking reciprocals, it follows that $\frac{N(t)}{t} \to 1/\mu$ w.p. 1.

Part (b) nearly follows from part (a), though there is a subtlety since $\{N(t)/t\}$ might be neither bounded nor monotone.
But it still follows from a *truncation argument*; see e.g. Grimmett & Stirzaker (1992), pp. 397–8, or Resnick (1992), p. 191.   ∎

Consider again Example (4.6.3), with $\{X_n\}_{n=0}^{\infty}$ an irreducible recurrent discrete Markov chain, and $N(t)$ the number of visits to state i by time t.
Then $\{N(t)\}$ is a renewal process. What is $\mu \equiv \mathbf{E}(Y_2)$?
Here $\mu = m_i =$ the *mean recurrence time* of state i, as in Section 2.8.
By the Elementary Renewal Theorem part (a), $N(t)/t \to 1/m_i$.
But here $N(n) = \#\{k \le n : X_k = i\} = \sum_{k=1}^{n} \mathbf{1}_{X_k=i}$.
Hence, $\lim_{n\to\infty} \frac{1}{n} \sum_{k=1}^{n} \mathbf{1}_{X_k=i} = 1/m_i$.
That is, the fraction of time the chain spends at i converges to $1/m_i$.
This fact was already shown previously in (2.8.3) above!
(Then, since $\lim_{n\to\infty} \frac{1}{n} \sum_{k=1}^{n} \mathbf{1}_{X_k=i} = \pi_i$, this showed $\pi_i = 1/m_i$.)
Furthermore, by the Elementary Renewal Theorem part (b), or by the Bounded Convergence Theorem since in this case $N(t) \le t$ so $N(t)/t \le 1$, we also have $\mathbf{E}[N(t)]/t \to 1/m_i$.

We next consider questions about $N(t+h) - N(t)$.
That is, about the number of arrivals between times t and $t+h$.
We are particular interesed in large t and fixed h. We have:

(4.6.5) Blackwell Renewal Theorem. Consider a renewal process as above, with $\mu := \mathbf{E}(Y_2) < \infty$. Assume that Y_2 is *non-arithmetic*, i.e. there is no $\lambda > 0$ with $\mathbf{P}(Y_2 = k\lambda$ for some $k \in \mathbf{Z}) = 1$ (similar to aperiodicity). Then for any fixed $h > 0$,

$$\lim_{t\to\infty} \mathbf{E}[N(t+h) - N(t)] = h/\mu \,.$$

This is a more "refined" theorem than (4.6.4), since it doesn't just consider the overall average $N(t)/t$, but rather considers the specific number of renewals between times t and $t+h$.
For a proof, see e.g. Grimmett & Stirzaker (1992), pp. 408–409, or Resnick (1992), Section 3.10.3.

In terms of the arrival times $T_1, T_2, \ldots$, we see that $N(t+h) - N(t) = \#\{i : t < T_i \le t+h\}$, so the Blackwell Renewal Theorem can also be written as $\lim_{t\to\infty} \mathbf{E}[\#\{i : t < T_i \le t+h\}] = h/\mu$.

What about $\mathbf{P}[N(t+h) - N(t) > 0]$, i.e. the probability that there is at least one renewal between times t and $t+h$?
This can also be written as $\mathbf{P}[\exists\, n : t < T_n \le t+h]$.
Well, if $h > 0$ is sufficiently small, then $N(t+h) - N(t)$ usually equals either 0 or 1, i.e. it is approximately an *indicator function*.
But for an indicator function, $\mathbf{P}(1_A > 0) = \mathbf{P}(A) = \mathbf{E}(1_A)$.
Hence, for small h, $\mathbf{P}[N(t+h) - N(t) > 0] \approx \mathbf{E}[N(t+h) - N(t)]$.
So, by the Blackwell Renewal Theorem, for small $h > 0$ and large t,

$$(4.6.6) \qquad \mathbf{P}[N(t+h) - N(t) > 0] \approx h/\mu .$$

In particular:

(4.6.7) Corollary. If h is small enough that $\mathbf{P}(Y_n < h) = 0$, then

$$\lim_{t\to\infty} \mathbf{P}[\exists\, n : t < T_n \le t+h] = \lim_{t\to\infty} \mathbf{P}[N(t+h) - N(t) \ge 1] = h/\mu .$$

Proof: There cannot be two renewals within time $[t, t+h]$, so

$$\lim_{t\to\infty} \mathbf{P}[N(t+h) - N(t) \ge 1] = \lim_{t\to\infty} \mathbf{E}[N(t+h) - N(t)] .$$

Hence, the result follows from the Blackwell Renewal Theorem. ∎

(4.6.8) Problem. Let $Y_1, Y_2, \ldots$ be i.i.d. $\sim$ Uniform[10, 20]. Let $T_n = Y_1 + Y_2 + \ldots + Y_n$, and let $N(t) = \max\{n \ge 0 : T_n < t\}$.
(a) Compute $\lim_{t\to\infty} [N(t)/t]$. **[sol]**
(b) Compute $\lim_{t\to\infty} \mathbf{P}(\exists\, n \ge 1 : t < T_n < t+6)$. **[sol]**

(4.6.9) Problem. Let $Y_1, Y_2, \ldots$ be i.i.d. $\sim$ Uniform[0, 10]. Let $T_0 = 0$, and $T_n = Y_1 + Y_2 + \ldots + Y_n$ for $n \ge 1$. Let $N(t) = \max\{n \ge 0 : T_n \le t\}$.
(a) Compute $\lim_{t\to\infty} [N(t)/t]$.
(b) Approximate $\mathbf{E}(\#\{n \ge 1 : 1234 < T_n < 1236\})$.
(c) Approximate $\mathbf{P}(\exists\, n \ge 1 : 1234 < T_n < 1236\})$.

(4.6.10) Problem. Let $\{Y_n\}_{n=0}^{\infty}$ be i.i.d., each having the density function $y^3/64$ for $0 < y < 4$, otherwise 0. Let $T_0 = 0$, and let $T_n = Y_1 + Y_2 + \ldots + Y_n$ for $n \geq 1$. Find (with explanation) a good approximation for $\mathbf{E}[\,\#\{n \geq 1 : 12345 < T_n < 12348\}\,]$.

Finally, suppose that at the kth renewal time T_k, you receive a certain *reward* (or cost) R_k.
Assume the $\{R_k\}$ are i.i.d., and are independent of $\{N(t)\}$.
Let $R(t) = \sum_{k=1}^{N(t)} R_k$ be the total reward received by time t.
Then $\{R(t)\}$ is a *Renewal Reward Processes*.
What is the average reward rate, i.e. $\lim_{t\to\infty}[R(t)/t]$? We have:

(4.6.11) Renewal Reward Theorem.   If $\{N(t)\}$ is a renewal process, with $\mu := \mathbf{E}(Y_2) < \infty$, then $\lim_{t\to\infty} \frac{R(t)}{t} = \frac{\mathbf{E}[R_1]}{\mu}$ w.p. 1.

Proof: Here

$$\frac{R(t)}{t} = \frac{\sum_{k=1}^{N(t)} R_k}{t} = \frac{\sum_{k=1}^{N(t)} R_k}{N(t)} \frac{N(t)}{t}.$$

As $t \to \infty$, w.p. 1, $\frac{\sum_{k=1}^{N(t)} R_k}{N(t)} \to \mathbf{E}[R_1]$ by the Law of Large Numbers (A.7.2), and $\frac{N(t)}{t} \to 1/\mu$ by the Elementary Renewal Theorem (4.6.4)(a).
Hence, $\frac{R(t)}{t} \to \mathbf{E}[R_1] \times (1/\mu) = \mathbf{E}[R_1]/\mu$, as claimed.   ∎

(4.6.12) Machine Purchases Example.

We illustrate the above ideas with an extended example about machine purchases. We make the following assumptions:
• You run an industrial factory which requires a certain machine.
• Each new machine's lifetime L has distribution Uniform[0,10] (in months), after which it breaks down.
• Your strategy is to buy a new machine as soon as your old machine breaks down, <u>or</u> after S months if the machine hasn't broken down.
• You get to <u>choose</u> any value S with $0 \leq S \leq 10$.
• A new machine costs 30 (in millions of dollars).

- If your machine breaks down before you replace it, that costs you an extra 5 (in millions of dollars).

Question: How <u>many</u> machines will you buy per month on average?

Solution: Here $\mu = \mathbf{E}(Y_2) = \mathbf{E}[\min(L, S)]$.
So, $\mu = S\,\mathbf{P}(L > S) + \mathbf{E}[L \mid L < S]\,\mathbf{P}(L < S) = S[(10 - S)/10] + (S/2)(S/10) = S - (S^2/20)$.
So, by the Elementary Renewal Theorem part (a), $\lim_{t\to\infty} N(t)/t = 1/\mu = 1/[S-(S^2/20)]$, i.e. you will buy an average of $1/[S-(S^2/20)]$ machines per month, or one machine every $[S - (S^2/20)]$ months.
e.g. if $S = 10$, then you will buy an average of 0.2 machines per month, i.e. one machine every 5 months. (Of course.)
Or, if $S = 9$, then you will buy an average of 0.202 machines per month, i.e. one machine every 4.95 months.
If $S = 5$: 0.267 machines per month, i.e. one every 3.75 months.
If $S = 2$: 0.556 machines per month, i.e. one every 1.8 months.
If $S = 1$: 1.053 machines per month, i.e. one every 0.95 months.

Question: As a function of S, about how many machines will you expect to buy between times 562 and 562.5 (in months)?

Solution: To apply the Blackwell Renewal Theorem, we want $t = 562$, and $t + h = 562.5$ so $h = 0.5$.
Hence, by the Blackwell Renewal Theorem, since $t = 562$ is reasonably large, the expected number of purchases between times 562 and 562.5 is approximately $h/\mu = 0.5/[S - (S^2/20)]$.
e.g. if $S = 9$, then this equals about 0.101.

Question: As a function of S, what is the approximate probability that you will buy <u>at least one</u> machine between times 562 and 562.5 (in months)?

Solution: Since $h = 0.5$ is quite small, the number of machines purchased in this interval will usually equal 0 or 1, so (4.6.6) applies. Hence, the probability of buying at least one machine during this time interval is approximately equal to the expected number of purchases, i.e. to $h/\mu = 0.5/[S - (S^2/20)]$.

Question: What is your long-run machine <u>cost</u> per month?

Solution: Let T_k be the time (in months) of the purchase of your k^{th} machine (with $T_0 = 0$).
Let $Y_k = T_k - T_{k-1}$ be the kth interarrival time.
Let $R(t)$ be your total machine cost by time t.
Then this is a renewal reward process!
Here $E[R_1] = 30 + 5\,\mathbf{P}(L < S) = 30 + 5\,(S/10) = 30 + S/2$.
And, again $\mu = \mathbf{E}[Y_2] = S - (S^2/20)$ as above.
Hence, w.p. 1, by the Renewal Reward Theorem,

$$\lim_{t \to \infty} \frac{R(t)}{t} = \frac{\mathbf{E}[R_1]}{\mu} = \frac{30 + S/2}{S - (S^2/20)} \, .$$

If $S = 10$ (i.e., never replace early), this equals $\frac{35}{5} = 7$.
(This makes sense: you spend 35 about every 5 months.)
If $S = 9$, this equals $\frac{34.5}{99/20} \doteq 6.970$. Less!
If $S = 5$, this equals about 8.667. Much more!
Or, if $S = 2$, then it's 17.222, while if $S = 1$, then it's 32.105.
(For a plot as a function of S, see Figure 20.)

Question: What choice of S minimises your long-run cost?

Solution: Optimise with calculus! Here

$$\frac{d(\text{cost per month})}{dS} = \frac{d}{dS}\left(\frac{30 + S/2}{S - (S^2/20)}\right)$$

$$= \frac{[S - (S^2/20)][1/2] - [30 + S/2][1 - S/10]}{[S - (S^2/20)]^2}\,.$$

This equals 0 iff:

$$[S - (S^2/20)][1/2] - [30 + S/2][1 - S/10] = 0$$

$$S^2/40 + 3S - 30 = 0$$

$$S^2 + 120S - 1200 = 0$$

$$S = \left(-120 \pm \sqrt{120^2 - 4(1)(-1200)}\right)\Big/(2(1))$$

$$S = -60 \pm \sqrt{4800} \doteq -129.282 \quad \text{or} \quad 9.282$$

So, the long-run cost is minimised when $S = -60 + \sqrt{4800} \doteq 9.282$.

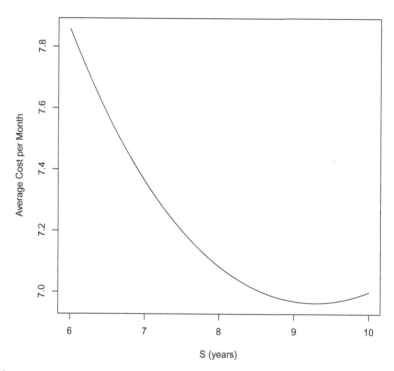

Figure 20: Average cost per month for Example (4.6.12).

So, your best policy is to buy a new machine as soon as your old machine breaks, or is 9.282 months old, whichever comes first.
Then the long-run cost per month is $\lim_{t\to\infty}[R(t)/t] \doteq 6.964 < 7$.
So, this is cheaper than always waiting until your machine breaks.

(4.6.13) Problem. Suppose you work at a job where after each payment, the boss waits an i.i.d. Uniform[3,7] amount of time, and then pays you an i.i.d. Uniform[10,20] number of dollars.
(a) Compute a good approximation to the expected number of payments that you will receive between times 133 and 135.
(b) Compute a good approximation to the probability that you will receive <u>at least one payment</u> between times 133 and 135.
(c) Let $R(t)$ be the total amount you get paid up to time t. Compute your long-run average rate of payment, i.e. $\lim_{t\to\infty}[R(t)/t]$.

(4.6.14) Problem. Let $\{X_n\}_{n=0}^{\infty}$ be a discrete-time irreducible Markov chain on the state space $S = \{1, 2, 3, 4\}$, with $X_0 = 4$, having stationary distribution $\pi = (1/7, 1/7, 2/7, 3/7)$. Let $T_1, T_2, \ldots$ be the return times to the state 4. Let $Z_1, Z_2, \ldots$ be i.i.d. $\sim$ Uniform$[-12, 2]$. Let $N(t) = \max\{m \geq 0 : T_m \leq t\}$, and let $H(t) = \sum_{k=1}^{N(t)} Z_k$. Compute (with explanation) $\lim_{t \to \infty} [H(t) / t]$.

4.7. Continuous-Space Markov Chains

In this book we have studied many stochastic processes.
For most of them, time and space were both <u>discrete</u>.
By contrast, Brownian motion had <u>continuous</u> time and space.
And, Poisson processes and Renewal processes and the Markov processes of Section 4.4 had continuous time but discrete space.
We now consider the final case, namely processes which have discrete time but continuous space (which is my main area of research!).

So, suppose that our state space S is <u>any</u> non-empty set, not necessarily discrete, e.g. $S = \mathbf{R}$.
But time is still discrete, i.e. the process still proceeds one step at a time as $X_0, X_1, X_2, \ldots$
The transition probabilities cannot be written p_{ij}, since the probability might be 0 of jumping to any one particular state.
Instead, we write the (one-step) transition probabilities from a state $x \in S$ as $P(x, A)$ for each (measurable) subset $A \subseteq S$.
Here $P(x, A)$ is the probability, if the chain is at a point x, that it will jump to <u>somewhere</u> within the subset A at the next step.
If S is finite or countable, then $P(x, \{j\})$ corresponds to the transition probability p_{xj} of a discrete Markov chain, as before.
Alternatively, for each fixed state $x \in S$, we can directly describe $P(x, \cdot)$ as a specific <u>probability distribution</u>.

We also require *initial probabilities*, which here are defined to be any probability distribution ν on S. (The simplest choice is to specify that X_0 equals some specific value, e.g. $X_0 = 5$.)

Then, generalising (1.2.2) and (4.4.1), we define a discrete-time, general state space (and time-homogeneous) *Markov chain* to be a sequence $X_0, X_1, X_2, \ldots$ of random variables on some (general) state space S, with

$$\mathbf{P}(X_0 \in A_0, \ X_1 \in A_1, \ldots, X_n \in A_n)$$

$$= \int_{x_0 \in A_0} \nu(dx_0) \int_{x_1 \in A_1} P(x_0, dx_1) \ldots$$

$$\ldots \int_{x_{n-1} \in A_{n-1}} P(x_{n-2}, dx_{n-1}) \int_{x_n \in A_n} P(x_{n-1}, dx_n),$$

for all $n \in \mathbf{N}$ and all (measurable) subsets $A_0, A_1, \ldots, A_n$.
If we want, we can simplify $\int_{x_n \in A_n} P(x_{n-1}, dx_n) = P(x_{n-1}, A_n)$.
But we <u>cannot</u> further simplify this expression; those "integrals" are
with respect to the (general) transition probabilities $P(x, \cdot)$ of the
Markov chain, and do not simplify in any easy way.
To make this more concrete, consider a simple example:

(4.7.1) Normal Example. Consider the Markov chain on the real
line (i.e. with $S = \mathbf{R}$) which begins at $X_0 = 4$, and has transition
probabilities $P(x, \cdot) = \text{Normal}(\frac{x}{2}, \frac{3}{4})$ for each $x \in S$.
That is, at each time $n \geq 1$, the Markov chain replaces the previous
value X_{n-1} with a new value X_n drawn from a normal distribution
with mean $X_{n-1}/2$ and variance $3/4$.
In other words, each time it divides the previous value by 2, and then
adds an independent Normal$(0, \frac{3}{4})$ random variable to it.
Equivalently, $X_n = \frac{1}{2} X_{n-1} + Z_n$, where the $\{Z_n\}$ are i.i.d. with
$Z_n \sim \text{Normal}(0, \frac{3}{4})$. (For a simulation, see Figure 21.)
Does this Markov chain have a stationary distribution?
Will it converge to it? (Answers are coming soon!)

How can we define "irreducible" for a continuous-space chain?
On a discrete space, it meant that we had a positive probability of
eventually hitting any given state.
But for a chain on a continuous space like Example (4.7.1), we might
have probability 0 of ever hitting a specific state.
So, instead, we define irreducibility in terms of eventually hitting any
subset A of positive size.
Specifically, it has to have $\phi(A) > 0$, where ϕ is a *measure*, i.e. a
certain formal description of the size of A.

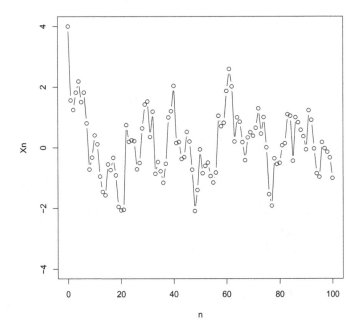

Figure 21: A simulation of the Normal Example (4.7.1).

(4.7.2) Definition. A Markov chain on a general state space S is ϕ-*irreducible* if there is some non-zero measure ϕ on S such that for all $x \in S$ and all subsets $A \subseteq S$ with $\phi(A) > 0$, the chain started at x has a positive probability of eventually hitting A.

The most common choice is where ϕ is *Lebesgue measure* on $\mathbf{R}$, corresponding to the *length* of intervals. Indeed, we have:

(4.7.3) Proposition. Suppose $S = \mathbf{R}$, and there is $\delta > 0$ such that for each $x \in S$, a Markov chain's transition probability distribution $P(x, \cdot)$ has positive density function at least on $[x - \delta, x + \delta]$. Then the chain is ϕ-irreducible, where ϕ is Lebesgue measure.

Proof: Let $x \in S$, and let A be any (measurable) subset of $\mathbf{R}$ with positive Lebesgue measure.
We need to show that, from x, the chain has positive probability of eventually hitting A.
Since A has positive measure, there must be some $m \in \mathbf{N}$ such that

$A \cap [-m, m]$ has positive measure.

Choose n large enough that $x - n\delta < -m$, and $x + n\delta > m$.

Since $P(x, \cdot)$ has positive density on $[x - \delta, x + \delta]$, the n-step transitions $P^n(x, \cdot)$ have positive density on $[x - n\delta, x + n\delta]$.

But $[x - n\delta, x + n\delta] \supseteq [-m, m] \supseteq A \cap [-m, m]$.

So, the chain has positive probability of hitting $A \cap [-m, m]$ after n steps. Hence, the chain is ϕ-irreducible, as claimed.   ∎

Next, how can we define "aperiodic" in continuous space?

On a discrete space, we defined it as the gcd of the possible return times to an individual state.

But for a chain on a continuous space like Example (4.7.1), we might have probability 0 of ever returning.

So, instead, we define aperiodicity by analogy to the *Cyclic Decomposition Lemma* as discussed in Problem (2.5.6):

(4.7.4) Definition. The *period* of a general-state-space Markov chain is the largest (finite) positive integer b for which there is a "cyclic" disjoint partition $S = S_0 \mathbin{\dot\cup} S_1 \mathbin{\dot\cup} \ldots \mathbin{\dot\cup} S_{b-1}$ such that if $i \in S_r$ for some $0 \le r \le b - 2$, then $P(x, S_{i+1}) = 1$ for all $x \in S_i$ $(1 \le i \le d - 1)$ and $P(x, S_1) = 1$ for all $x \in S_d$. (That is, the chain is forced to repeatedly move from S_0 to S_1 to S_2 to $\ldots$ to S_{b-1} and then back to S_0.)

(4.7.5) Definition. A general-state-space Markov chain is *aperiodic* if its period from Definition (4.7.4) is equal to 1.

Then, similar to Proposition (4.7.3), we have:

(4.7.6) Proposition. Suppose there is $\delta > 0$ such that for each $x \in S = \mathbf{R}$, a Markov chain's transition probability distribution $P(x, \cdot)$ has positive density function at least on $[x - \delta, x + \delta]$. Then the chain is aperiodic.

Proof: Suppose to the contrary that the period is $b \ge 2$.

Since $S = S_0 \cup S_1 \cup \ldots \cup S_{b-1}$, there must be at least one subset S_i with positive Lebesgue measure. Choose any $x \in S_i$.

Then it is only possible to move from x to S_i in multiples of b steps.

Find $m \in \mathbf{N}$ such that $S_i \cap [-m, m]$ has positive measure.

Since $P(x, \cdot)$ has positive density on $[x - \delta, x + \delta]$, the n-step transitions $P^n(x, \cdot)$ have positive density on $[x - n\delta, x + n\delta]$.
Choose n_0 large enough that $x - n_0\delta < -m$, and $x + n_0\delta > m$.
Then, from x, the chain has positive probability of hitting $S_i \cap [-m, m]$ after n steps for any $n \geq n_0$, including both for $n = n_0$ and for $n = n_0 + 1$.
But if $b \geq 2$, then n_0 and $n_0 + 1$ cannot both be multiples of b.
This contradicts the assumption that $b \geq 2$.
So, the chain must be aperiodic.   ∎

So what about *stationary* distributions?
In the discrete case, we defined this as $\pi_j = \sum_{i \in S} \pi_i p_{ij}$ for all j.
On a general state space, the corresponding definition is:

(4.7.7) Definition. A general-state-space Markov chain with transition probabilities $P(x, \cdot)$ on a state space S has *stationary* distribution π if $\pi(A) = \int_S \pi(dx) P(x, A)$ for all subsets $A \subseteq S$.

Equivalently, just like in the discrete case, π is stationary iff it has the property that whenever we begin with $X_0 \sim \pi$, then also $X_1 \sim \pi$. As in the discrete case, Markov chains on general state spaces may or may not have stationary distributions.

If $S = \mathbf{R}$ and the transition probabilities have density functions, i.e. $P(x, A) = \int_A f(x, y)\, dy$ for some fixed function f, then a density function π is stationary iff $\int_{\mathbf{R}} \pi(x) f(x, y)\, dx = \pi(y)$ for all $y \in \mathbf{R}$. As in Proposition (2.2.2), this follows if the chain is *reversible*, i.e.

$$(4.7.8) \qquad \pi(x)\, f(x, y) \;=\; \pi(y)\, f(y, x), \qquad x, y \in S.$$

For the Normal Example (4.7.1), let $\pi = \text{Normal}(0, 1)$ be the standard normal distribution.
Recall that $X_n = \frac{X_{n-1}}{2} + Z_n$, where $\{Z_n\}$ are i.i.d. $\sim \text{Normal}(0, \frac{3}{4})$.
So, if $X_0 \sim \pi = \text{Normal}(0, 1)$, then $\frac{X_0}{2} \sim \text{Normal}(0, \frac{1}{4})$, and hence by the normal sum property (A.3.13),

$$X_1 \;=\; \frac{X_0}{2} + Z_1 \;\sim\; \text{Normal}\Big(0, \frac{1}{4}\Big) + \text{Normal}\Big(0, \frac{3}{4}\Big)$$

$$= \text{Normal}(0, 1) \;=\; \pi.$$

So, π is a stationary distribution for this chain!

We can now state (similar to Theorem (2.4.1)):

(4.7.9) General Space Convergence Theorem. If a discrete-time, general-state-space Markov chain is ϕ-irreducible, aperiodic, and has stationary distribution π, then for "π-a.e." $x \in S$,

$$\lim_{n \to \infty} \mathbf{P}_x(X_n \in A) = \pi(A), \qquad A \subseteq S.$$

For a proof of this result, see e.g. Roberts and Rosenthal (2004), or Meyn and Tweedie (1997).

Here, "π-a.e." means from *almost every* choice of initial state x, i.e. the theorem might fail[15] from certain x but they all have π measure equal to zero; see Problem (4.7.13).

Consider again the Normal Example (4.7.1).

We saw above that it has stationary distribution $\pi = $ Normal$(0, 1)$. And, since $P(x, \cdot) = $ Normal$(\frac{x}{2}, \frac{3}{4})$ has positive density everywhere, it satisfies the conditions of Propositions (4.7.3) and (4.7.6) for any $\delta > 0$ (even $+\infty$), so it is ϕ-irreducible and aperiodic.

Hence, by Theorem (4.7.9), for this example, $\lim_{n \to \infty} \mathbf{P}(X_n \in A) = \pi(A)$ for all A. In particular, for any $a < b$,

$$\lim_{n \to \infty} \mathbf{P}(a < X_n < b) = \Phi(b) - \Phi(a),$$

where Φ is as in (A.3.12).

(4.7.10) Problem. Consider a discrete-time Markov chain with state space $S = \mathbf{R}$, and with transition probabilities such that $P(x, \cdot)$ is uniform on the interval $[x-5, x+5]$. Determine whether or not this chain is ϕ-irreducible, and whether or not it is aperiodic. [sol]

(4.7.11) Problem. Consider a discrete-time Markov chain with state space $S = \mathbf{R}$, and transition probabilities such that $P(x, \cdot) = $

---

[15]This issue is related to a property called *Harris recurrence*; for more details see e.g. G.O. Roberts and J.S. Rosenthal (2006), "Harris Recurrence of Metropolis-Within-Gibbs and Trans-Dimensional Markov Chains", *Annals of Applied Probability* **16**, 2123–2139.

Uniform$[0, 1]$ for all x. (In particular, $P(x, A)$ does not depend on x.)

(a) Is this chain ϕ-irreducible? [**sol**]

(b) Is this chain aperiodic? [**sol**]

(c) Does this chain have a stationary distribution π? [**sol**]

(d) If yes, does $\lim_{n\to\infty} \mathbf{P}_x(X_n \in A) = \pi(A)$ for all $A \subseteq S$ and π-a.e. $x \in S$? [**sol**]

(4.7.12) Problem. (*) **(a)** Prove that a Markov chain on a <u>countable</u> state space S is ϕ-irreducible iff there is $j \in S$ with $f_{ij} > 0$ for all $i \in S$, i.e. j can be reached from any state i.

(b) Give an example of a Markov chain on a countable state space which is ϕ-irreducible, but which is <u>not</u> irreducible according to our previous (discrete state space) definition.

(4.7.13) Problem. (*) Consider a Markov chain on the discrete state space $S = \{1, 2, 3, \ldots\}$, with transition probabilities $p_{1,1} = 1$, and $p_{x,1} = 1/x^2$ and $P_{x,x+1} = 1 - (1/x^2)$ for $x \geq 2$.

(a) Let $\pi\{1\} = 1$, and $\pi\{x\} = 0$ for all $x \geq 2$. Prove that π is a stationary distribution for this chain.

(b) Prove that this chain is ϕ-irreducible, for the choice $\phi = \pi$.

(c) Prove this chain is aperiodic. [Hint: Here $P(1, \{1\}) > 0$.]

(d) Prove that when $x = 1$, $\lim_{n\to\infty} \mathbf{P}_1(X_n = 1) = 1$.

(e) Prove that if $X_0 = x \geq 2$, $\mathbf{P}_x[X_n = x + n$ for all $n] = \prod_{n=x}^{\infty}(1 - (1/n^2))$, an infinite product.

(f) Prove that if $X_0 = x \geq 2$, then $\mathbf{P}_x[X_n = x + n$ for all $n] > 0$. [Hint: You may use without proof the fact that if $0 \leq c_n < 1$ with $\sum_{n=1}^{\infty} c_n < \infty$, then $\prod_{n=1}^{\infty}(1 - c_n) > 0$.]

(g) Prove that for $x \geq 2$, $\lim_{n\to\infty} \mathbf{P}_x(X_n = 1)$ does <u>not</u> equal $\pi\{1\}$.

(h) Explain why this does not contradict Theorem (4.7.9).

4.8. Application – MCMC Algorithms (Continuous)

In Section 2.6, we discussed the very important *Markov chain Monte Carlo (MCMC)* algorithms, especially the *Metropolis algorithm*, in the case of <u>discrete</u> distributions.

But many of the most important applications of MCMC, e.g. statistical models for medical research, are continuous.

Now that we have studied continuous-state-space Markov chains, we can understand the Metropolis algorithm in the continuous case, too.

To begin, let π be a density function on, say, $S = \mathbf{R}$. As before, we want to *sample* from π, i.e. create random variables X with $\mathbf{P}(X = i) \approx \pi_i$ for all $i \in S$. We will again create a Markov chain $X_0, X_1, X_2, \ldots$ such that $\lim_{n\to\infty} \mathbf{P}(X_n \in A) = \pi(A) := \int_A \pi(x)\,dx$ for all subsets $A \subseteq S$. This is, again, the *Markov chain Monte Carlo (MCMC)* method.

We use a general *Metropolis algorithm* (related to Problem 2.6.4). For each $x \in S$, let $q(x, \cdot)$ be a *proposal density* function on S. Assume that q is *symmetric*, i.e. $q(x, y) = q(y, x)$ for all $x, y \in S$. Then, define transition probabilities $P(x, \cdot)$ for a Markov chain on S such that for $y \neq x$, $P(x, \cdot)$ has density $q(x, y) \min[1, \frac{\pi(y)}{\pi(x)}]$. (As a special case, if $\pi(x) = 0$, we always take $\frac{\pi(y)}{\pi(x)} = 1$.) To make $P(x, \cdot)$ have total probability 1, we also let

$$P(x, \{x\}) = 1 - \int_S q(x, y) \min\left[1, \frac{\pi(y)}{\pi(x)}\right] dy.$$

That is, $P(x, \cdot)$ <u>mostly</u> has a density, except that it also has a positive probability of staying exactly where it is (i.e., at x). The Markov chain first chooses X_0 from any initial distribution ν, e.g. perhaps it just sets $X_0 = 5$. Then, for $n = 1, 2, \ldots$, it generates X_n from $P(X_{n-1}, \cdot)$ as above. An equivalent algorithmic version is: Given X_{n-1}, choose $Y_n \sim q(X_{n-1}, \cdot)$, and let $U_n \sim \text{Uniform}[0, 1]$ be i.i.d., and then let

$$X_n = \begin{cases} Y_n, & U_n \leq \frac{\pi(Y_n)}{\pi(X_{n-1})} & (\text{``accept''}) \\ X_{n-1}, & \text{otherwise} & (\text{``reject''}) \end{cases}$$

which gives the same transition probabilities by (A.3.9).

(4.8.1) Continuous MCMC Convergence Theorem. Assume that $\pi(x) > 0$ for all $x \in S$, and there is $\delta > 0$ such that $q(x, y) > 0$ whenever $|y - x| \leq \delta$. Then the Metropolis algorithm converges in distribution to π, i.e. the above Markov chain has the property that

$\lim_{n\to\infty} \mathbf{P}_x[X_n \in A] = \pi(A) := \int_A \pi(z)\,dz$ for all subsets $A \subseteq S$ and for π-a.e. initial state $x \in S$.

Proof: For $y \neq x$, let $f(x,y) = \pi(x)\,q(x,y)\,\min[1, \frac{\pi(y)}{\pi(x)}]$ be the density for $P(x,\cdot)$ at y.
Then since $q(x,y) = q(y,x)$, we compute that

$$\pi(x)\,f(x,y) \;=\; \pi(x)\,q(x,y)\,\min[1, \frac{\pi(y)}{\pi(x)}] \;=\; q(x,y)\,\min[\pi(x),\,\pi(y)]$$

$$=\; q(y,x)\,\min[\pi(x),\,\pi(y)] \;=\; \pi(y)\,f(y,x)\,.$$

So, the chain is <u>reversible</u> w.r.t. π as in (4.7.8).
So, π is a <u>stationary distribution</u> for the chain.
Also, since $\pi(x) > 0$ and $q(x,y) > 0$ for $|y - x| \leq \delta$, it follows that $P(x,\cdot)$ has positive density on $[x - \delta, x + \delta]$.
Hence, by Proposition (4.7.3), the chain is ϕ-irreducible.
And, by Proposition (4.7.6), the chain is aperiodic.
Hence, the result follows from Theorem (4.7.9).   ∎

(4.8.2) Problem. Let $\pi(x) = \frac{1}{2}e^{-|x|}$ and $q(x,y) = 1(|y - x| \leq 1/2)$ for $x, y \in \mathbf{R}$. Describe explicitly the steps of a Metropolis algorithm for this π and q. [sol]

(4.8.3) Problem. Let $\pi(x) = c_1 e^{-5x^2}$ and $q(x,y) = c_2 e^{-3(y-x)^2}$ for $x, y \in \mathbf{R}$ and normalising constants $c_1, c_2 > 0$. Describe explicitly the steps of a Metropolis algorithm for this π and q.

(4.8.4) Problem. Let π be the density of the Uniform$[-8, 8]$ distribution, and $q(x, \cdot)$ be the density of Uniform$[x - 2, x + 2]$.
(a) Describe explicitly the steps of a Metropolis algorithm for this π and q (including what happens if $\pi(X_{n-1}) = 0$).
(b) Prove this chain is ϕ-irreducible and aperiodic. [Hint: Do Propositions (4.7.3) and (4.7.6) apply? Can similar reasoning be used?]
(c) Prove this chain converges in distribution to π.

(4.8.5) Problem. (*) [*Continuous Metropolis-Hastings.*]   Let S, π, and q be as in this section, with $\pi(x) > 0$ and $q(x,y) > 0$ whenever $|y - x| \leq \delta$ for some $\delta > 0$, except do <u>not</u> assume that q is symmetric, merely that it is symmetrically positive, i.e. $q(x, y) > 0$ iff $q(y, x) > 0$. For $x \in S$, let $P(x, \cdot)$ have density for $y \neq x$ given by $q(x, y) \min[1, \frac{\pi(y) \, q(y,x)}{\pi(x) \, q(x,y)}]$, and let

$$P(x, \{x\}) \;=\; 1 - \int_S q(x, y) \, \min\left[1, \, \frac{\pi(y) \, q(y, x)}{\pi(x) \, q(x, y)}\right] dy .$$

Prove that for this Markov chain, $\lim_{n \to \infty} \mathbf{P}_x[X_n \in A] = \pi(A)$ for all $A \subseteq S$ and π-a.e. $x \in S$. [Hint: Follow the proof of Theorem (4.8.1).]

A. Appendix: Background

Before reading this book, you should already have a solid background in undergraduate-level probability theory, including discrete and continuous probability distributions, random variables, joint distributions, expected values, inequalities, conditional probability, limit theorems, etc.

You should also know basic calculus concepts such as derivatives and integrals, and basic real analysis including limits and infinite series (but <u>not</u> including measure theory), and such linear algebra notions as matrix multiplication and eigenvectors.

This Appendix reviews certain background material and supplementary results that will be required and referenced at various points in this book.

A.1. Notation

The following notation will be used throughout this book:

$\mathbf{Z}$ are the integers, $\mathbf{N}$ are the positive integers, and $\mathbf{R}$ are the real numbers.

$\mathbf{P}(\cdots)$ is the *probability* of an event.

$\mathbf{E}(\cdots)$ is the *expected value* of a random variable.

X, Y, etc. usually denote *random variables*, which take on different values with different probabilities.

$\{X_n\}_{n=0}^{\infty}$ is an infinite sequence of random variables, often a *stochastic process* such as a *Markov chain* or a *martingale*.

$\mathbf{P}_i(\cdots)$ and $\mathbf{E}_i(\cdots)$ are *shorthand notation* for probability and expectation, respectively, conditional on $X_0 = i$, i.e. *assuming* that the stochastic process $\{X_n\}$ begins in the state i.

$\mathbf{1}_{\cdots}$ is the *indicator function* of an event, e.g. $\mathbf{1}_{X>5}$ equals 1 if $X > 5$, or equals 0 if $X \leq 5$. It follows that $\mathbf{E}(\mathbf{1}_A) = \mathbf{P}(A)$.

iff means "if and only if", i.e. the statements before and after must be either both true or both false.

$\exists$ means "there exists", and $\forall$ means "for all".

w.r.t. is an occasional shorthand for "with respect to".

$\emptyset$ is the *empty set*, i.e. the set containing no elements.

inf is the *infimum* of a set (a generalisation of *minimum*), with the convention that $\inf(\emptyset) = \infty$.

sup is the *supremum* of a set (a generalisation of *maximum*), with the convention that $\sup(\emptyset) = -\infty$.

$\cup$ is set *union*, e.g. $A \cup B$ is the set of all elements in either A or B (or both).

$\cap$ is set *intersection*, e.g. $A \cap B$ is the set of all elements in both A and B.

$\setminus$ is *set difference*, e.g. $A \setminus B$ is the set of all elements in A but not B.

$\dot{\cup}$ is *disjoint union*, e.g. $A \dot{\cup} B$ is the union of A and B, where A and B are *disjoint* (i.e. $A \cap B = \emptyset$).

A.2. Basic Probability Theory

The reader is assumed to already know undergraduate-level probability theory. Here we just mention a few basic facts.

If X is *discrete* (i.e. takes on only a finite or countable number of different values), its *expected value* is $\mathbf{E}(X) = \sum_{\ell} \ell \, \mathbf{P}(X = \ell)$. If X is non-negative-integer-valued, $\mathbf{E}(X) = \sum_{\ell=1}^{\infty} \ell \, \mathbf{P}(X = \ell)$.

If X is *absolutely continuous* with *density function* f, then $\mathbf{P}(a < X < b) = \int_a^b f(x) \, dx$ for any $a < b$, and $\mathbf{E}(X) = \int_{-\infty}^{\infty} x \, f(x) \, dx$.

If $\mathbf{E}(X) = m$, then the *variance* of X is $v = \mathbf{E}[(X - m)^2] = \mathbf{E}(X^2) - m^2$.

One trick for computing expectations is:

(A.2.1) Proposition. If Z is non-negative-integer-valued, then

$$\mathbf{E}(Z) = \sum_{k=1}^{\infty} \mathbf{P}(Z \geq k).$$

Proof: This follows since

$$\sum_{k=1}^{\infty} \mathbf{P}(Z \geq k) = \sum_{k=1}^{\infty} [\mathbf{P}(Z = k) + \mathbf{P}(Z = k+1) + \mathbf{P}(Z = k = 2) + \ldots]$$

$$= [\mathbf{P}(Z = 1) + \mathbf{P}(Z = 2) + \mathbf{P}(Z = 3) + \mathbf{P}(Z = 4) + \ldots]$$
$$+ [\mathbf{P}(Z = 2) + \mathbf{P}(Z = 3) + \mathbf{P}(Z = 4) + \ldots]$$

$$+ \left[\mathbf{P}(Z = 3) + \mathbf{P}(Z = 4) + \mathbf{P}(Z = 5) + \ldots \right]$$
$$= \mathbf{P}(Z = 1) + 2 \, \mathbf{P}(Z = 2) + 3 \, \mathbf{P}(Z = 3) + \ldots$$
$$= \sum_{\ell=1}^{\infty} \ell \, \mathbf{P}(Z = \ell) \; = \; \mathbf{E}(Z) . \qquad \blacksquare$$

Probabilities satisfy *additivity* over countable disjoint subsets: if A and B are *disjoint* (i.e. $A \cap B = \emptyset$), then $\mathbf{P}(A \cup B) = \mathbf{P}(A) + \mathbf{P}(B)$. More generally, if $A_1, A_2, \ldots$ are a disjoint sequence, then

$$(\mathbf{A.2.2}) \qquad \mathbf{P}\left(\bigcup_n A_n \right) \; = \; \sum_n \mathbf{P}(A_n) .$$

For example, if X and Y are discrete random variables, then $\{X = i, \ Y = j\}$ is disjoint, and $\bigcup_j \{X = i, \ Y = j\} = \{X = i\}$, so

$$(\mathbf{A.2.3}) \qquad \mathbf{P}(X = i) \; = \; \sum_j \mathbf{P}(X = i, \ Y = j) ,$$

which is one version of the *Law of Total Probability*.

It follows that probabilities are *monotone*, i.e.:

$$(\mathbf{A.2.4}) \qquad \text{If } A \subseteq B, \ \text{then } \mathbf{P}(A) \le \mathbf{P}(B) .$$

A.3. Standard Probability Distributions

Here we review a few standard probability distributions that will arise in this book, that you should already know.

If S is a non-empty finite set, the *(discrete) uniform distribution* on S gives equal probability to each point in S. Hence, if $X \sim \text{Uniform}(S)$, then

$$(\mathbf{A.3.1}) \qquad \mathbf{P}(X = s) \; = \; \frac{1}{|S|} \quad \text{for each } s \in S ,$$

where $|S|$ is the number of elements in S.

The *binomial distribution* Binomial(n, p) represents the probabilities for the number of successes from n independent trials each with probability p of succeeding. Hence, if $X \sim \text{Binomial}(n, p)$, then for $k = 0, 1, \ldots, n$,

$$\mathbf{P}(X = k) = \binom{n}{k} p^k (1 - p)^{n-k},$$

i.e.

(**A.3.2**) $$\mathbf{P}(X = k) = \frac{n!}{k!(n - k)!} p^k (1 - p)^{n-k},$$

where $n! = n(n - 1)(n - 2) \ldots (2)(1)$ is n *factorial*.

The *geometric distribution* Geometric(p) represents the probabilities for the number of trials before the first success in an infinite sequence of independent trials each having success probability p. Hence, if $X \sim \text{Geometric}(p)$, then

(**A.3.3**) $$\mathbf{P}(X = k) = p^k (1 - p), \quad \text{for } k = 0, 1, 2, \ldots,$$

and it is computed that $\mathbf{E}(X) = \frac{1}{p} - 1$. Alternatively, the geometric distribution is sometimes defined as one <u>more</u> than this (i.e., counting the first success in addition to the failures), so

(**A.3.4**) $$\mathbf{P}(X = k) = p^{k-1} (1 - p), \quad \text{for } k = 1, 2, 3, \ldots,$$

and $\mathbf{E}(X) = \frac{1}{p}$. We will use (A.3.3) or (A.3.4), as needed.

The *Poisson distribution* Poisson(λ) is surprisingly useful (e.g. when constructing a Poisson process). If $X \sim \text{Poisson}(\lambda)$, then

(**A.3.5**) $$\mathbf{P}(X = k) = \frac{1}{k!} \lambda^k e^{-\lambda}, \quad \text{for } k = 0, 1, 2, \ldots,$$

and

(**A.3.6**) $$\mathbf{E}(X) = \mathbf{Var}(X) = \lambda.$$

If $X \sim \text{Poisson}(\lambda_1)$ and $Y \sim \text{Poisson}(\lambda_2)$ are independent, then

(**A.3.7**) $$X + Y \sim \text{Poisson}(\lambda_1 + \lambda_2).$$

The *(continuous) uniform distribution* on an interval $[L, R]$ (where $L < R$) gives probability $(b - a)/(R - L)$ to $[a, b]$ whenever $L \leq a \leq$

$b \leq R$, with mean $(a+b)/2$. For example, if $X \sim \text{Uniform}[0,1]$, then $\mathbf{E}(X) = 1/2$, and

(A.3.8) $\mathbf{P}\left(\dfrac{1}{2} \leq X \leq \dfrac{2}{3}\right) = \dfrac{1}{2} - \dfrac{2}{3} = \dfrac{1}{6}$,   etc.

In particular, setting $a = 0$ and $b = y$,

(A.3.9) $\mathbf{P}(X \leq y) = \min(1, y)$,   for any $y \geq 0$.

The *exponential distribution Exponential*(λ) with *rate* $\lambda > 0$ has density $f(x) = \lambda e^{-\lambda x}$ for $x > 0$. If $X \sim \text{Exponential}(\lambda)$, then

(A.3.10) $\mathbf{P}(a < X < b) = \displaystyle\int_a^b \lambda e^{-\lambda x}\, dx$,

and $\mathbf{P}(X > c) = e^{-\lambda c}$ for any $c > 0$, and $\mathbf{E}(X) = 1/\lambda$. (In some other treatments, λ is instead replaced by $1/\lambda$ throughout.)

The *normal distribution Normal*(m, v) with mean m and variance v has density $f(x) = \frac{1}{\sqrt{2\pi v}} e^{-(x-m)^2/2v}$. In particular, the *standard normal distribution* $\text{Normal}(0,1)$ has density

(A.3.11) $\phi(x) = \dfrac{1}{\sqrt{2\pi}} e^{-x^2/2}$,

so if $Z \sim \text{Normal}(0,1)$, then

(A.3.12) $\Phi(u) := \mathbf{P}[Z \leq u] = \displaystyle\int_{-\infty}^{u} \dfrac{1}{\sqrt{2\pi}} e^{-x^2/2}\, dx$.

If $X \sim \text{Normal}(m_1, v_1)$ and $Y \sim \text{Normal}(m_2, v_2)$ are independent, then

(A.3.13) $X + Y \sim \text{Normal}(m_1 + m_2,\ v_1 + v_2)$.

A.4. Infinite Series and Limits

The infinite sum $\sum_{n=1}^{\infty} x_n$ is shorthand for $\lim_{N\to\infty} \sum_{n=1}^{N} x_n$.

If the x_n are non-negative, then $\sum_{n=1}^{\infty} x_n$ either (a) converges to some finite value, in which case we write $\sum_{n=1}^{\infty} x_n < \infty$, or (b) diverges to infinity, in which case we write $\sum_{n=1}^{\infty} x_n = \infty$.

For $c \in \mathbf{R}$ and $|r| < 1$, the *geometric series* is

(**A.4.1**)
$$\sum_{n=0}^{\infty} c\, r^n = c + cr + cr^2 + cr^3 + \ldots = \frac{c}{1-r}.$$

Since $\int_2^{\infty} t^{-a} dt < \sum_{n=2}^{\infty} n^{-a} < \int_1^{\infty} t^{-a} dt$, it follows that:

(**A.4.2**)
$$\sum_{n=1}^{\infty} (1/n^a) = \infty \qquad \underline{\text{iff}} \qquad a \leq 1.$$

We also have:

(**A.4.3**) If $x_n \geq 0$, and $\sum_{n=1}^{\infty} x_n < \infty$, then $\lim_{n \to \infty} x_n = 0$.

Proof: If $\sum_{n=1}^{\infty} x_n$ equals some finite number a, then

$$\lim_{n \to \infty} x_n = \lim_{n \to \infty} \left[\left(\sum_{i=1}^{n} x_i \right) - \left(\sum_{i=1}^{n-1} x_i \right) \right]$$

$$= \lim_{n \to \infty} \left(\sum_{i=1}^{n} x_i \right) - \lim_{n \to \infty} \left(\sum_{i=1}^{n-1} x_i \right) = a - a = 0. \qquad \blacksquare$$

The converse to (A.4.3) is false. For example, if $x_n = 1/n$, then $\lim_{n \to \infty} x_n = 0$, but $\sum_{n=1}^{\infty} x_n = \infty$ by (A.4.2). Hence:

(**A.4.4**) It is possible that $\sum_{n=1}^{\infty} x_n = \infty$ even if $\lim_{n \to \infty} x_n = 0$.

(**A.4.5**) **Problem.** Prove that $\sum_{n=2}^{\infty} (1 / [n \log(n)]) = \infty$.
[Hint: First show $\sum_{n=2}^{\infty} (1 / [n \log(n)]) \geq \int_2^{\infty} (dx / [x \log(x)]).$]

(**A.4.6**) **Problem.** Prove that $\sum_{n=3}^{\infty} (1 / [n \log(n) \log \log(n)]) = \infty$, but $\sum_{n=3}^{\infty} \left(1 / [n \log(n)[\log (\log(n))]^2] \right) < \infty$.

Next, we consider the *Cesàro sum*, i.e. the limit of partial averages $\lim_{n \to \infty} \frac{1}{n} \sum_{i=1}^{n} x_i$, and its relationship to $\lim_{n \to \infty} x_n$.

For example, suppose $x_n = 1$ for n odd, and $x_n = 0$ for n even. Then $\lim_{n \to \infty} x_n$ does not exist, but $\lim_{n \to \infty} \frac{1}{n} \sum_{i=1}^{n} x_i$ does exist (and is equal to $1/2$).

However, the reverse implication is always true, i.e. if the limit exists then the Cesàro sum also exists and equals the same value:

$$(\textbf{A.4.7}) \qquad \text{If } \lim_{n\to\infty} x_n = r, \quad \text{then} \quad \lim_{n\to\infty} \frac{1}{n} \sum_{i=1}^{n} x_i = r.$$

The *lim inf (limit infimum)* and *lim sup (limit supremum)* are respectively the smallest and largest possible limits of a sequence. A limit exists iff its lim inf and lim sup are equal.

One simple but helpful fact about limits is the *Sandwich Theorem* (or, *Squeeze Theorem*):

(A.4.8) If $\{a_n\}$, $\{b_n\}$, and $\{c_n\}$ are three sequences with $a_n \le b_n \le c_n$, and if $\lim_{n\to\infty} a_n = \lim_{n\to\infty} c_n = L$, then $\lim_{n\to\infty} b_n = L$.

Limits can be used to define *order notation*. A quantity is $O(\cdot)$ if, in a certain limit, it is bounded above by a finite constant times "$\cdot$". For example, as $h \searrow 0$, the quantities h^2 and $5h^2$ and $16h^3 + 12h^4$ are all $O(h)$, since e.g. $\lim_{h\searrow 0} \frac{5h^2}{h^2} < \infty$ and $\lim_{h\searrow 0} \frac{16h^3+12h^4}{h^2} < \infty$. In fact, $16h^3 + 12h^4$ is $o(h)$, where the small o means that $\lim_{h\searrow 0} \frac{16h^3+12h^4}{h^2} = 0$.

For example, from a first-order Taylor series expansion,

$$(\textbf{A.4.9}) \qquad e^x = 1 + x + O(x^2), \qquad \text{as } x \to 0.$$

A.5. Bounded, Finite, and Infinite

X is a *bounded random variable* if there is $M < \infty$ with $\textbf{P}(|X| \le M) = 1$, i.e. if X is always in some interval $[-M, M]$ for some finite number M.

X is a *finite random variable* if $\textbf{P}(|X| < \infty) = 1$, i.e. if $\textbf{P}(|X| = \infty) = 0$, i.e. if X always takes on finite values.

A random variable X has *finite expectation* if $\textbf{E}|X| < \infty$; this is also sometimes called being *integrable*.

If a random variable is bounded then it must have finite expectation; indeed in that case $\textbf{E}|X| \le M$.

But the converse is false, e.g. if $\textbf{P}(X = 2^{k/2}) = 2^{-k}$ for $k = 1, 2, 3, \ldots$, then $\textbf{P}(|X| > M) > 0$ for any $M < \infty$, but by (A.4.1) we have $\mathbf{E}|X| = \sum_{k=1}^{\infty} 2^{k/2}(2^{-k}) = \sum_{k=1}^{\infty} 2^{-k/2} = 2^{-1/2}/(1 - 2^{-1/2}) \doteq 2.414 < \infty$.

If a random variable has finite expectation then it must be finite. Indeed, $\mathbf{E}|X| = \sum_{\ell} \ell \, \mathbf{P}(|X| = \ell) \geq (\infty) \, \mathbf{P}(|X| = \infty)$, which shows:

(A.5.1) If $\mathbf{P}(|X| = \infty) > 0$, then $\mathbf{E}|X| = \infty$.

Hence, also:

(A.5.2) If $\mathbf{E}|X| < \infty$, then $\mathbf{P}(|X| = \infty) = 0$, i.e. $\mathbf{P}(|X| < \infty) = 1$.

But the converse is false, e.g. if $\mathbf{P}(X = 2^k) = 2^{-k}$ for $k = 1, 2, 3, \ldots$, then $\mathbf{P}(|X| < \infty) = 1$ so X is finite, but $\mathbf{E}|X| = \sum_{k=1}^{\infty} 2^k (2^{-k}) = \sum_{k=1}^{\infty} (1) = \infty$.

A.6. Conditioning

If $\mathbf{P}(B) > 0$, then the *conditional probability* of A given B is

$$(\mathbf{A.6.1}) \qquad \mathbf{P}(A \mid B) \;=\; \frac{\mathbf{P}(A \cap B)}{\mathbf{P}(B)},$$

from which it follows that

$$(\mathbf{A.6.2}) \qquad \mathbf{P}(A \cap B) \;=\; \mathbf{P}(B)\,\mathbf{P}(A \mid B).$$

If Y is a discrete random variable, and $\mathbf{P}(Y = y) > 0$, then

$$\mathbf{P}(A \mid Y = y) \;=\; \frac{\mathbf{P}(A, \, Y = y)}{\mathbf{P}(Y = y)}.$$

If X is also discrete, with finite expectation, then we define its *conditional expectation* as

$$\mathbf{E}(X \mid A) \;=\; \sum_{x} x \, \mathbf{P}(X = x \mid A).$$

So, if X and Y are both discrete, and $\mathbf{P}(Y = y) > 0$, then

$$(\mathbf{A.6.3}) \qquad \mathbf{E}(X \mid Y = y) \;=\; \sum_{x} x \, \mathbf{P}(X = x \mid Y = y).$$

(A.6.4) Law of Total Expectation. If X and Y are discrete random variables, then

$$\mathbf{E}(X) = \sum_y \mathbf{P}(Y = y)\, \mathbf{E}(X \mid Y = y),$$

i.e. we can compute $\mathbf{E}(X)$ by averaging conditional expectations.

Proof: We compute that

$$\sum_y \mathbf{P}(Y = y)\, \mathbf{E}(X \mid Y = y)$$

$$= \sum_y \mathbf{P}(Y = y) \sum_x x\, \mathbf{P}(X = x \mid Y = y)$$

$$= \sum_y \mathbf{P}(Y = y) \sum_x x\, \frac{\mathbf{P}(X = x,\, Y = y)}{\mathbf{P}(Y = y)}$$

$$= \sum_x x \sum_y \mathbf{P}(X = x,\, Y = y)$$

$$= \sum_x x\, \mathbf{P}(X = x) = \mathbf{E}(X)$$

where the last line uses the Law of Total Probability (A.2.3). ∎

If $X = 1_A$ is an indicator function, then (A.6.4) becomes:

(A.6.5) $$\mathbf{P}(A) = \sum_y \mathbf{P}(Y = y)\, \mathbf{P}(A \mid Y = y).$$

We sometimes consider $\mathbf{E}(X \mid Y)$ to be another random variable, the *conditional expectation* of X given Y, whose value when $Y = y$ is equal to $\mathbf{E}(X \mid Y = y)$ as above. We then have:

(A.6.6) Double-Expectation Formula. $\mathbf{E}\big[\mathbf{E}(X \mid Y)\big] = \mathbf{E}(X)$, i.e. the random variable $\mathbf{E}(X \mid Y)$ equals X on average.

Proof: Since $\mathbf{E}(X \mid Y)$ is equal to $\mathbf{E}(X \mid Y = y)$ with probability $Y = y$, we compute that

$$\mathbf{E}\big[\mathbf{E}(X \mid Y)\big] = \sum_y \mathbf{P}(Y = y)\, \mathbf{E}(X \mid Y = y),$$

whence the result follows from (A.6.4).   ∎

Another useful fact is:

(A.6.7) Conditional Factoring. If $h(Y)$ is a function of Y,

$$\mathbf{E}[h(Y)\,X\,|\,Y] \;=\; h(Y)\,\mathbf{E}[X\,|\,Y]\,,$$

i.e. when conditioning on Y, we can treat any function of Y as a *constant* and factor it out.

Proof: This follows since

$$\mathbf{E}[h(Y)\,X\,|\,Y=y] \;=\; \sum_r r\,\mathbf{P}[h(Y)\,X = r\,|\,Y=y]$$

$$= \; \sum_x [h(y)\,x]\,\mathbf{P}[X = x\,|\,Y=y]$$

$$= \; h(y)\sum_x x\,\mathbf{P}[X = x\,|\,Y=y]$$

$$= \; h(y)\,\mathbf{E}[X\,|\,Y=y]\,.   \blacksquare$$

In particular, setting $X = 1$, we obtain that

$$(\mathbf{A.6.8}) \qquad\qquad \mathbf{E}[h(Y)\,|\,Y] \;=\; h(Y)\,,$$

i.e. conditioning on Y has no effect on a function of Y.

If X and Y are general (not discrete) random variables, then a formal definition of $\mathbf{E}(X\,|\,Y)$ is more technical, involving σ-algebras (see e.g. Section 13 of Rosenthal, 2006, or many other measure-theoretic probability books). Intuitively, $\mathbf{E}(X\,|\,Y)$ is still the average value of X conditional on the value of Y. And, the Double-Expectation Formula (A.6.6) and conditional factoring (A.6.7) still hold.

Finally, conditioning gives us the *memoryless property* (or, *forgetfulness property*) of the Exponential distribution:

(A.6.9) Proposition. If $X \sim$ Exponential(λ), and $a, b > 0$, then $\mathbf{P}(X > b + a\,|\,X > a) = \mathbf{P}(X > b)$. (So, if X is a waiting time, then the probability you will need to wait a more minutes is unaffected by knowing that you've already waited b minutes.)

Proof: We compute that

$$\mathbf{P}(X > b + a \mid X > a) \; = \; \frac{\mathbf{P}(X > b + a, \; X > a)}{\mathbf{P}(X > a)}$$

$$= \; \frac{\mathbf{P}(X > b + a)}{\mathbf{P}(X > a)} \; = \; \frac{e^{-\lambda(b+a)}}{e^{-\lambda a}} \; = \; e^{-\lambda b} \; = \; \mathbf{P}(X > b) \, . \qquad \blacksquare$$

A.7. Convergence of Random Variables

There are various senses in which a sequence $X_1, X_2, X_3, \ldots$ of random variables could converge to a random variable X:

Convergence in distribution means that for any $a < b$, $\lim_{n \to \infty} \mathbf{P}(a < X_n < b) = \mathbf{P}(a < X < b)$.

Weak convergence means $\forall \epsilon > 0$, $\lim_{n \to \infty} \mathbf{P}(|X_n - X| \geq \epsilon) = 0$.

Strong convergence (or, convergence *with probability 1* or *w.p. 1*) means $\mathbf{P}(\lim_{n \to \infty} X_n = X) = 1$, i.e. the random sequence of values always converges.

The random variable X is often just a constant, e.g. $X = 0$.

Strong convergence implies weak convergence, and weak convergence implies convergence in distribution; for proofs see e.g. Propositions 5.2.3 and 10.2.1 of Rosenthal (2006) or many other advanced probability books.

However, the converse is false. For example, suppose the $\{X_n\}$ are mostly equal to 0, except that one of $X_1, \ldots, X_9$ (chosen uniformly at random) equals 1, and one of $X_{10}, \ldots, X_{99}$ equals 1, and one of $X_{100}, \ldots, X_{999}$ equals 1, etc. This sequence has an infinite number of 1's, so $\{X_n\}$ does not converge strongly to 0. But $\mathbf{P}(X_n \neq 0) \to 0$, so $\{X_n\}$ does converge weakly to 0.

Convergence to infinity requires slightly modifying the definitions. We say a sequence $\{X_n\}$ of random variables *converges weakly to positive infinity* if for all finite K,

$$(\mathbf{A.7.1}) \qquad\qquad \lim_{n \to \infty} \mathbf{P}(X_n < K) \; = \; 0 \, .$$

(For negative infinity, replace $X_n < K$ by $X_n > K$.)

The *Law of Large Numbers* (LLN) says that if the sequence $\{X_n\}$ is *i.i.d.* (i.e., both *independent* and *identically distributed*), with common finite mean m, then the sequence $\frac{1}{n}\sum_{i=1}^{n} X_i$ converges to m (both weakly and strongly), i.e.

$$(\textbf{A.7.2}) \qquad \lim_{n\to\infty} \frac{1}{n}\sum_{i=1}^{n} X_i = m \quad \text{w.p. } 1.$$

The *Central Limit Theorem* (CLT) says that if the sequence $\{X_n\}$ is *i.i.d.*, with common mean m and common finite variance v, then the sequence $\frac{1}{\sqrt{nv}}\sum_{i=1}^{n}(X_i - m)$ converges in distribution to the Normal$(0,1)$ distribution, so for any $a < b$,

$$(\textbf{A.7.3}) \qquad \lim_{n\to\infty} \mathbf{P}\left(a < \frac{1}{\sqrt{nv}}\sum_{i=1}^{n}(X_i - m) < b\right) = \int_a^b \phi(x)\, dx,$$

where $\phi(x) = \frac{1}{\sqrt{2\pi}} e^{-x^2/2}$ is again the Normal$(0,1)$ density.

A.8. Continuity of Probabilities

Suppose $A_1, A_2, \ldots$ are a sequence of events which "increase" to an event A, meaning that $A_1 \subseteq A_2 \subseteq A_3 \subseteq \ldots$, and $\bigcup_n A_n = A$. Then we can write $\{A_n\} \nearrow A$, and think of A as a sort of "limit" of the events $\{A_n\}$, i.e. $\lim_{n\to\infty} A_n = A$.

But here $A = \bigcup_n (A_n \setminus A_{n-1})$ (taking $A_0 = \emptyset$), which is a disjoint union, so $\mathbf{P}(A) = \sum_n \left(\mathbf{P}(A_n) - \mathbf{P}(A_{n-1})\right) = \lim_{n\to\infty}\left(\mathbf{P}(A_n) - \mathbf{P}(A_0)\right) = \lim_{n\to\infty} \mathbf{P}(A_n)$, i.e. probabilities respect these limits:

$$(\textbf{A.8.1}) \qquad \text{If } \{A_n\} \nearrow A, \text{ then } \lim_{n\to\infty} \mathbf{P}(A_n) = \mathbf{P}(A).$$

Similarly, if $A_1 \supseteq A_2 \supseteq A_3 \supseteq \ldots$, and $\bigcap_n A_n = A$, then we can write $\{A_n\} \searrow A$. Taking complements, $\{A_n^C\} \nearrow A^C$, so by (A.8.1) we have $\lim_{n\to\infty} \mathbf{P}(A_n^C) = \mathbf{P}(A^C)$, so $\lim_{n\to\infty}\left(1 - \mathbf{P}(A_n^C)\right) = 1 - \mathbf{P}(A^C)$, and hence:

$$(\textbf{A.8.2}) \qquad \text{If } \{A_n\} \searrow A, \text{ then } \lim_{n\to\infty} \mathbf{P}(A_n) = \mathbf{P}(A).$$

These facts are called *continuity of probabilities*; see e.g. Section 3.3 of Rosenthal (2006), or many other advanced probability books.

If X is a random variable, and A_n is the event $\{X \geq n\}$, and A is the event $\{X = \infty\}$, then $\{A_n\} \searrow A$, so $\lim_{n\to\infty} \mathbf{P}(A_n) = \mathbf{P}(A)$, i.e.

$$(\textbf{A.8.3}) \qquad \lim_{n\to\infty} \mathbf{P}(X \geq n) = \mathbf{P}(X = \infty).$$

A.9. Exchanging Sums and Expectations

For a *finite* collection of random variables, the expectation of a sum always equals the sum of the expectations, e.g. $\mathbf{E}(X + Y) = \mathbf{E}(X) + \mathbf{E}(Y)$, and more generally

$$(\textbf{A.9.1}) \qquad \mathbf{E}\left(\sum_{i=1}^{M} X_i\right) = \sum_{i=1}^{M} \mathbf{E}(X_i).$$

That is, expectation satisfies finite *linearity*.

But for an *infinite* collection, this might not be true. For example, suppose Z is a random variable with $\mathbf{P}(Z = n) = 2^{-n}$ for all $n \in \mathbf{N}$, and

$$Y_n = \begin{cases} 2^n, & Z = n \\ -2^{n+1}, & Z = n+1 \\ 0, & \text{otherwise} \end{cases}$$

i.e. $Y_n = 2^n \, \mathbf{1}_{Z=n} - 2^{n+1} \, \mathbf{1}_{Z=n+1}$ for each $n \in \mathbf{N}$.

Then $\mathbf{E}(Y_n) = 2^n \, \mathbf{P}(Z = n) + (-2^{n+1}) \, \mathbf{P}(Z = n+1) = 2^n \, (2^{-n}) - 2^{n+1}(2^{-(n+1)}) = 1 - 1 = 0$, so also $\sum_{n=1}^{\infty} \mathbf{E}(Y_n) = 0$. However,

$$\sum_{n=1}^{\infty} Y_n = \sum_{n=1}^{\infty} (2^n \, \mathbf{1}_{Z=n} - 2^{n+1} \, \mathbf{1}_{Z=n+1})$$

$$= (2^1 \, \mathbf{1}_{Z=1} - 2^2 \, \mathbf{1}_{Z=2}) + (2^2 \, \mathbf{1}_{Z=2} - 2^3 \, \mathbf{1}_{Z=3}) + (2^3 \, \mathbf{1}_{Z=3} - 2^4 \, \mathbf{1}_{Z=4}) + \dots$$

$$= 2 \, \mathbf{1}_{Z=1}$$

since the rest of the sum all cancels out. Hence, $\mathbf{E}[\sum_{n=1}^{\infty} Y_n] = \mathbf{E}[2 \, \mathbf{1}_{Z=1}] = 2 \, \mathbf{P}(Z = 1) = 2 \, (1/2) = 1$. So, in this case, $\sum_{n=1}^{\infty} \mathbf{E}(Y_n) = 0 \neq 1 = \mathbf{E}[\sum_{n=1}^{\infty} Y_n]$, i.e. we cannot interchange the sum and expected value.

What goes wrong in this example? Well, this expression really equals $\infty - \infty$. That is, both the positive and the negative parts of the expression are infinite, so the calculations involve an infinite amount of cancellation, and the final result depends on the order in which the cancelling is done.

However, if the Y_n are *non-negative* random variables, then there is no cancellation, so such problems do not arise:

(A.9.2) Countable Linearity. If $\{Y_n\}$ is a sequence of non-negative random variables, then $\sum_{n=1}^{\infty} \mathbf{E}(Y_n) = \mathbf{E}[\sum_{n=1}^{\infty} Y_n]$, i.e. we can exchange the order of the sum and expected values.

Indeed, this follows from the Monotone Convergence Theorem (below) upon setting $X_n = \sum_{k=1}^{n} Y_k$.

Similarly, upon replacing expected values by sums, it follows that if x_{nk} are non-negative real numbers, then

$$(\text{A.9.3}) \qquad \sum_n \sum_k x_{nk} = \sum_k \sum_n x_{nk}.$$

A.10. Exchanging Expectations and Limits

If a sequence of random variables $\{X_n\}$ converges to X with probability 1 (cf. Section A.7), this might not imply that $\lim_{n \to \infty} \mathbf{E}(X_n) = \mathbf{E}(X)$.

For example, if $U \sim \text{Uniform}(0, 1)$, and $X_n = n \, \mathbf{1}_{0 < U < 1/n}$, then $X_n \to 0$ (since for any value of U, we have $X_n = 0$ for all $n > 1/U$), but $\mathbf{E}(X_n) = (n)(1/n) = 1$ for all n. So, $\mathbf{E}[\lim_{n \to \infty} X_n] = \mathbf{E}[0] = 0$, but $\lim_{n \to \infty} \mathbf{E}[X_n] = \mathbf{E}[1] = 1$, i.e. we cannot interchange the limit and expectation.

However, under some *additional* conditions, if $X_n \to X$ w.p. 1 then this does imply that $\lim_{n \to \infty} \mathbf{E}(X_n) = \mathbf{E}(X)$. Indeed, the following results are well-known (for proofs see e.g. Section 9.1 of Rosenthal, 2006, or many other advanced probability books).

(A.10.1) Bounded Convergence Theorem. If $X_n \to X$ w.p. 1, and the $\{X_n\}$ are uniformly bounded (i.e., there is $M < \infty$ with $|X_n| \leq M$ for all n), then $\lim_{n \to \infty} \mathbf{E}(X_n) = \mathbf{E}(X)$.

(A.10.2) Monotone Convergence Theorem. If $X_n \to X$ w.p. 1, and $0 \leq X_1 \leq X_2 \leq X_3 \leq \ldots$, then $\lim_{n\to\infty} \mathbf{E}(X_n) = \mathbf{E}(X)$.

(A.10.3) Dominated Convergence Theorem. If $X_n \to X$ w.p. 1, and there is some random variable Y with $\mathbf{E}|Y| < \infty$ and $|X_n| \leq Y$ for all n, then $\lim_{n\to\infty} \mathbf{E}(X_n) = \mathbf{E}(X)$.

A.11. Exchanging Limits and Sums

We next ask when we can exchange limits and sums, i.e. be sure that $\lim_{n\to\infty} \sum_{k=1}^{\infty} x_{nk} = \sum_{k=1}^{\infty} \lim_{n\to\infty} x_{nk}$.

If there are only a *finite* number of k in the sum, then this always holds, since (from first-year calculus) the limit of finite sums always equals the sum of the corresponding limits.

But if there are an *infinite* number of different k, then this might fail. For example, suppose $x_{nn} = 1$ but $x_{nk} = 0$ for $k \neq n$. Then $\lim_{n\to\infty} x_{nk} = 0$ for each fixed k (since $x_{nk} = 0$ whenever $n > k$), so $\sum_{k=1}^{\infty} \lim_{n\to\infty} x_{nk} = \sum_{k=1}^{\infty} 0 = 0$. On the other hand, $\sum_k x_{nk} = 1$ for each fixed n, so $\lim_{n\to\infty} \sum_{k=1}^{\infty} x_{nk} = \lim_{n\to\infty} 1 = 1$. Hence, in this case, $\lim_{n\to\infty} \sum_{k=1}^{\infty} x_{nk} = 1 \neq 0 = \sum_{k=1}^{\infty} \lim_{n\to\infty} x_{nk}$.

Still, the following result gives a useful condition under which the limit and sum can be exchanged. It may be regarded as a special case of the *Weierstrasse M-test*, or of the above *Dominated Convergence Theorem* (with expected values replaced by sums); here we simply call it the *M-test*.

(A.11.1) M-test. Let $\{x_{nk}\}_{n,k\in\mathbf{N}}$ be real numbers. Suppose that $\lim_{n\to\infty} x_{nk}$ exists for each fixed $k \in \mathbf{N}$, and that $\sum_{k=1}^{\infty} \sup_n |x_{nk}| < \infty$. Then $\lim_{n\to\infty} \sum_{k=1}^{\infty} x_{nk} = \sum_{k=1}^{\infty} \lim_{n\to\infty} x_{nk}$.

Proof: Let $a_k = \lim_{n\to\infty} x_{nk}$. The assumption implies that $\sum_{k=1}^{\infty} |a_k| < \infty$. Hence, by replacing x_{nk} by $x_{nk} - a_k$, it suffices to assume for simplicity that $a_k = 0$ for all k.

Fix $\epsilon > 0$. Since $\sum_{k=1}^{\infty} \sup_n x_{nk} < \infty$, we can find $K \in \mathbf{N}$ such that $\sum_{k=K+1}^{\infty} \sup_n x_{nk} < \epsilon/2$. Since $\lim_{n\to\infty} x_{nk} = 0$, we can find (for $k = 1, 2, \ldots, K$) numbers N_k with $x_{nk} < \epsilon/2K$ for all $n \geq N_k$. Let $N = \max(N_1, \ldots, N_K)$. Then for $n \geq N$, we have $\sum_{k=1}^{\infty} x_{nk} <$

$K\frac{\epsilon}{2K} + \frac{\epsilon}{2} = \epsilon$. Since this is true for all $\epsilon > 0$, we must actually have $\lim_{n\to\infty} \sum_{k=1}^{\infty} x_{nk} \leq 0 = \sum_{k=1}^{\infty} a_k$.

Similarly, $\lim_{n\to\infty} \sum_{k=1}^{\infty} x_{nk} \geq \sum_{k=1}^{\infty} a_k$. The result follows. ∎

If $x_{nk} \geq 0$, and $\lim_{n\to\infty} x_{nk} = a_k$ for each fixed $k \in \mathbf{N}$, but we do *not* know that $\sum_{k=1}^{\infty} \sup_n x_{nk} < \infty$, then we still have

$$(\mathbf{A.11.2}) \qquad \lim_{n\to\infty} \sum_{k=1}^{\infty} x_{nk} \geq \sum_{k=1}^{\infty} a_k,$$

assuming this limit exists. Indeed, if not then we could find some finite $K \in \mathbf{N}$ with $\lim_{n\to\infty} \sum_{k=1}^{\infty} x_{nk} < \sum_{k=1}^{K} a_k$, contradicting the fact that $\lim_{n\to\infty} \sum_{k=1}^{\infty} x_{nk} \geq \lim_{n\to\infty} \sum_{k=1}^{K} x_{nk} = \sum_{k=1}^{K} a_k$.

A.12. Basic Linear Algebra

We first recall *matrix multiplication*. If A is an $r \times m$ matrix, and B is an $m \times n$ matrix, then the product AB is a $r \times n$ matrix whose ij entry (for $i = 1, 2, \ldots, r$ and $j = 1, 2, \ldots, n$) is:

$$(\mathbf{A.12.1}) \qquad (AB)_{ij} = \sum_{k=1}^{m} A_{ik} B_{kj}.$$

For example,

$$\begin{pmatrix} 1 & 2 & 3 & 4 \\ 5 & 6 & 7 & 8 \\ 9 & 10 & 11 & 12 \end{pmatrix} \begin{pmatrix} 13 & 14 \\ 15 & 16 \\ 17 & 18 \\ 19 & 20 \end{pmatrix} = \begin{pmatrix} 170 & 180 \\ 426 & 452 \\ 682 & 724 \end{pmatrix}$$

since e.g. $170 = 1 \times 13 + 2 \times 15 + 3 \times 17 + 4 \times 19$, and $452 = 5 \times 14 + 6 \times 16 + 7 \times 18 + 8 \times 20$, etc.

These formulas still hold for infinite matrices, where $m = \infty$, in which case the sum (A.12.1) becomes an infinite sum.

And, the $m \times m$ *identity matrix* has 1 on the diagonal, otherwise 0.

Of particular use in this book is when $r = 1$ and $m = n$, e.g.

$$(1 \quad 2 \quad 3) \begin{pmatrix} 4 & 5 & 6 \\ 7 & 8 & 9 \\ 10 & 11 & 12 \end{pmatrix} = (48 \quad 54 \quad 60).$$

In that case, if v is a $1 \times m$ vector, and A is an $m \times m$ matrix, then v is called a *left eigenvector* for A with *eigenvalue* λ if

$$(\textbf{A.12.2}) \qquad\qquad v\,A \;=\; \lambda\,v\,,$$

i.e. if multiplying v by the matrix A is equivalent[16] to just multiplying it by the scalar constant λ. In particular, if v is a left eigenvector with eigenvalue $\lambda = 1$, then $v\,A = v$ (which will correspond to a *stationary distribution*).

A.13. Miscellaneous Math Facts

First, the *triangle inequality* says $|a + b| \le |a| + |b|$, and more generally

$$(\textbf{A.13.1}) \qquad\qquad \left|\sum_i a_i\right| \;\le\; \sum_i |a_i|\,.$$

In terms of expected values, this implies that

$$(\textbf{A.13.2}) \qquad\qquad \left|\mathbf{E}(X)\right| \;\le\; \mathbf{E}|X|\,.$$

Second, the following approximation from Real Analysis helps with the factorials in the choose formula for the binomial distribution (A.3.2):

(A.13.3) Stirling's Approximation. If n is large, then $n! \approx (n/e)^n \sqrt{2\pi n}$. More precisely,

$$\lim_{n \to \infty} \frac{n!}{(n/e)^n \sqrt{2\pi n}} \;=\; 1\,.$$

Finally, to deal with aperiodicity, we need the following result from Number Theory about a property of the *greatest common divisor (gcd)*; for a proof, see e.g. Rosenthal (2006) p. 92, or Durrett (2011) p. 24.

---

[16]Note that in probability theory we usually put the vector v on the *left* of the matrix, though in pure linear algebra it is usually put on the right, and $v\,A$ is <u>not</u> the same as $A\,v$. (However, there is a connection since $(v\,A)^t = A^t\,v^t$ where "t" means transpose.)

(A.13.4) Number Theory Lemma. If a set A of positive integers is non-empty, and satisfies *additivity* (i.e. $m+n \in A$ whenever $m \in A$ and $n \in A$), and $gcd(A) = 1$, then there is some $n_0 \in \mathbf{N}$ such that $n \in A$ for all $n \geq n_0$, i.e. the set A includes all of the integers n_0, $n_0 + 1$, $n_0 + 2$, ...

B. Bibliography for Further Reading

The following works provide additional details and perspectives about stochastic processes (and have influenced this book in various ways), and might be useful for further reading:

P. Billingsley (1995), *Probability and Measure*, 3rd ed. John Wiley & Sons, New York. (Advanced.)

J.K. Blitzstein and J. Hwang (2019), *Introduction to Probability*, 2nd ed. Taylor & Francis Group. (Elementary; see especially Chapters 11 and 12.) `http://probabilitybook.net/`

R. Douc, E. Moulines, P. Priouret, and P. Soulier (2018), *Markov Chains*. Springer, New York. (Advanced.)

L.E. Dubins and L.J. Savage (2014), *How to Gamble if you Must: Inequalities for Stochastic Processes*. Dover Publications, New York, reprint edition.

R. Durrett (2011), *Essentials of Stochastic Processes*, 2nd ed. Springer, New York. `http://www.math.duke.edu/~rtd/EOSP/`

B. Fristedt and L. Gray (1997), *A Modern Approach to Probability Theory*. Birkhauser, Boston.

G.R. Grimmett and D.R. Stirzaker (1992), *Probability and Random Processes*, 2nd ed. Oxford University Press.

O. Häggström (2002), *Finite Markov Chains and Algorithmic Applications*. Cambridge University Press.

S. Karlin and H. M. Taylor (1975), *A First Course in Stochastic Processes*. Academic Press, New York.

G.E. Lawler (2006), *Introduction to Stochastic Processes*, 2nd ed. Chapman & Hall.

S.P. Meyn and R.L. Tweedie (1993), *Markov Chains and Stochastic Stability*. Springer-Verlag, London. (Advanced.) `http://probability.ca/MT/`

S. Resnick (1992), *Adventures in Stochastic Processes*. Birkhäuser, Boston.

G.O. Roberts and J.S. Rosenthal (2004), "General state space Markov chains and MCMC algorithms", *Probability Surveys* 1, 20–71. `https://www.i-journals.org/ps/viewarticle.php?id=15`

J.S. Rosenthal (1995), "Convergence rates of Markov chains", *SIAM Review* **37**, 387–405.

J.S. Rosenthal (2006), *A First Look at Rigorous Probability Theory*, 2nd ed. World Scientific Publishing Company, Singapore. (Especially Chapters 7, 8, 14, and 15.)

S.M. Ross (2014), *Introduction to Probability Models*, 11th ed. Academic Press, Oxford.

D. Williams (1991), *Probability with Martingales.* Cambridge University Press. (Advanced.)

C. Solutions to Problems Marked [sol]

1.1.1: *To get to pad #1 after 2 steps, the frog could either first jump from pad #20 to pad #1 and then stay there on the second jump with probability* $(1/3)(1/3)$, *or it could first stay at pad #20 and then jump from there to pad #1 with probability* $(1/3)(1/3)$. *So,* $\mathbf{P}$(*at pad #1 after 2 steps*) $= (1/3)(1/3) + (1/3)(1/3) = 2/9$.

1.2.3(a): $\mathbf{P}(X_0 = 20,\ X_1 = 19) = \nu_{20}p_{20,19} = (1)(1/3) = 1/3$.

1.2.3(b): $\mathbf{P}(X_0 = 20,\ X_1 = 19,\ X_2 = 20) = \nu_{20}p_{20,19}p_{19,20} = (1)(1/3)(1/3) = 1/9$.

1.3.3(a):

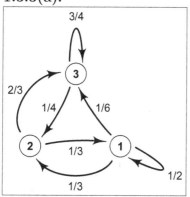

1.4.1(a): $\mathbf{P}(X_1 = 3) = \mu_3^{(1)} = \sum_{i \in S} \nu_i p_{i3} = \nu_1 p_{13} + \nu_2 p_{23} + \nu_3 p_{33} = (1/7)(1/2) + (2/7)(1/3) + (4/7)(1/2) = 19/42$.

1.4.6(a):

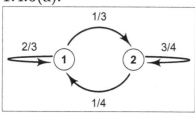

1.4.6(b): $p_{12}^{(2)} = \sum_{k \in S} p_{1k}p_{k2} = p_{11}p_{12} + p_{12}p_{22} = (2/3)(1/3) + (1/3)(3/4) = 2/9 + 1/4 = 17/36$.

1.4.7: *Ehrenfest's Urn adds or subtracts 1 each time, so the only ways to get from state 1 to state 3 in 3 steps are* $1 \to 0 \to$

$1 \to 2$, $1 \to 2 \to 1 \to 2$, and $1 \to 2 \to 3 \to 2$. The first has probability $(1/5)(1)(4/5) = 4/25 = 20/125$, the second has probability $(4/5)(2/5)(4/5) = 32/125$, and the third has probability $(4/5)(3/5)(3/5) = 36/125$, so $p_{12}^{(3)} = (20/125) + (32/125) + (36/125) = 88/125$.

1.5.4: From state 1, if the chain doesn't hit state 2 right away then it never will, so $f_{12} = p_{12} = 1/4$. However, whether it jumps first to 2 or to 3, it must eventually jump to 3, so $f_{13} = 1$.

1.6.18(a): Yes, it's irreducible: since $p_{ij} > 0$ for all $i, j \in S$, therefore $i \to j$ for all $i, j \in S$.

1.6.18(b): Yes. Since the chain is irreducible, and has a finite state space, it is recurrent by the Finite Space Theorem. Hence, $f_{ii} = 1$ for all $i \in S$. In particular, $f_{11} = 1$.

1.6.18(c): Yes. Since $f_{11} = 1$ and $1 \to 2$ (by irreducibility), it follows from the f-Lemma (with $i = 2$ and $j = 1$) that $f_{21} = 1$. (Or compute it directly.)

1.6.18(d): Yes it is. By the Recurrent State Theorem, since $f_{11} = 1$, therefore $\sum_{n=1}^{\infty} p_{11}^{(n)} = \infty$. (Or, use the Finite Space Theorem.)

1.6.18(e): Yes it is. By the Recurrence Equivalences Theorem, since the chain is irreducible and $f_{11} = 1$, therefore $\sum_{n=1}^{\infty} p_{ij}^{(n)} = \infty$ for all $i, j \in S$, and in particular $\sum_{n=1}^{\infty} p_{21}^{(n)} = \infty$.

1.6.19(a): No, it isn't: since $p_{21} = 0$ and $p_{31} = 0$, it is impossible to ever get from state 2 to state 1, so the chain is <u>not</u> irreducible.

1.6.19(b): $f_{11} = p_{11} = 1/2$, since once we leave state 1 then we can never return to it. But $f_{22} = 1$, since from state 2 we will either return to state 2 immediately, or go to state 3 but from there eventually return to state 2. Similarly, $f_{33} = 1$, since from state 3 we will either return to state 3 immediately, or go to state 2 but from there eventually return to state 3.

1.6.19(c): Since $f_{11} = 1/2 < 1$, therefore state 1 is transient. But since $f_{22} = f_{33} = 1$, therefore states 2 and 3 are recurrent.

1.6.19(d): From state 1, the chain might stay at state 1 for some number of steps, but with probability 1 will eventually move to state 2. Then, from state 2, the chain might stay at state 2 for some number of steps, but with probability 1 will eventually move to state 3. So, $f_{13} = 1$.

1.6.20(a): *Does not exist. We know from the Cases Theorem that for any irreducible transient Markov chain,* $\sum_{n=1}^{\infty} p_{k\ell}^{(n)} < \infty$ *for all* $k, \ell \in S$. *In particular, we must have* $\lim_{n \to \infty} p_{k\ell}^{(n)} = 0$. *So, it is impossible that* $p_{k\ell}^{(n)} \geq 1/3$ *for all* $n \in \mathbf{N}$.

1.6.20(b): *Does not exist. We know from the Recurrence Equivalences Theorem that for any irreducible Markov chain, if* $\sum_{n=1}^{\infty} p_{k\ell}^{(n)} = \infty$ *for any one pair* $k, \ell \in S$, *then the chain is recurrent, and* $f_{ij} = 1$ *for all* $i, j \in S$.

1.6.20(c): *Exists. For example, let* $S = \{1, 2, 3\}$, *with* $p_{12} = 1$, $p_{22} = 1/3$, $p_{23} = 2/3$, *and* $p_{33} = 1$ *(with* $p_{ij} = 0$ *otherwise). Then if the chain is started at* $k = 1$, *then it will initially follow the path* $1 \to 2 \to 2 \to 2 \to 2 \to 2 \to 3$ *with probability* $(1)(1/3)(1/3)(1/3)(1/3)(2/3) > 0$, *after which it will remain in state 3 forever.*

1.6.20(d): *Exists. For example, consider simple random walk with* $p = 3/4$, *so* $S = \mathbf{Z}$ *and* $p_{i,i+1} = 3/4$ *and* $p_{i,i-1} = 1/4$ *for all* $i \in S$ *(with* $p_{ij} = 0$ *otherwise). Let* $k = 0$ *and* $\ell = 5$. *Then by (1.6.17),* $f_{05} = 1$, *and the chain is irreducible and transient. (Of course,* S *is infinite here; if* S *is finite then all irreducible chains are recurrent.)*

1.6.22(b): $p_{32}^{(2)} = \sum_{k \in S} p_{3k} p_{k2} = p_{31} p_{12} + p_{32} p_{22} + p_{33} p_{32} + p_{34} p_{42} = (0)(1/2) + (1/7)(2/3) + (2/7)(1/7) + (4/7)(0) = 2/21 + 2/49 = 20/147$.

1.6.22(c): *The subset* $C = \{1, 2\}$ *is* <u>closed</u> *since* $p_{ij} = 0$ *for* $i \in C$ *and* $j \notin C$. *Furthermore, the Markov chain restricted to* C *is irreducible (since it's possible to go* $1 \to 2 \to 1$*), and* C *is finite. Hence, by the Finite Space Theorem, we must have* $\sum_{n=1}^{\infty} p_{12}^{(n)} = \infty$.

1.6.22(d): *Here* $f_{32} = \sum_{n=1}^{\infty} \mathbf{P}_3[\text{first hit 2 at time } n] = \sum_{n=1}^{\infty} (2/7)^{n-1}(1/7) = (1/7)/(1 - (2/7)) = (1/7)/(5/7) = 1/5$. *Or, alternatively,* $f_{32} = \mathbf{P}_3[\text{hit 2 when we first leave 3}] = \mathbf{P}_3[\text{hit 2} \mid \text{leave 3}] = (1/7)/((1/7) + (4/7)) = 1/5$. *Or, alternatively, by the* f-Expansion, $f_{32} = p_{32} + p_{31}f_{12} + p_{33}f_{32} + p_{34}f_{42} = (1/7) + 0 + (2/7)f_{32} + 0$, so $(5/7)f_{32} = 1/7$, so $f_{32} = (1/7)/(5/7) = 1/5$.

1.7.8: $\mathbf{P}(T = \infty)$ is equivalent to the probability that Gambler's Ruin with $p = 3/5$ starting at state 2 will <u>never</u> reach state 0, which is one minus the probability it will <u>ever</u> reach state 0. Hence, from $(1.7.5)$, $\mathbf{P}(T = \infty) = 1 - (\frac{p}{1-p})^{-2} = 1 - (3/2)^{-2} = 1 - 4/9 = 5/9$.

2.1.2(a):

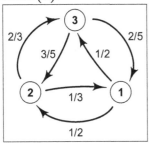

2.1.2(b): $p_{11}^{(2)} = \sum_{j \in S} p_{1j}p_{j1} = p_{11}p_{11} + p_{12}p_{21} + p_{13}p_{31} = (0)(0) + (1/2)(1/3) + (1/2)(2/5) = (1/6) + (1/5) = 11/30$.

2.1.2(c): We need $\pi P = \pi$, i.e. $\sum_{i \in S} \pi_i p_{ij} = \pi_j$ for all $j \in S$. When $j = 1$ this gives $\pi_2(1/3) + \pi_3(2/5) = \pi_1$. When $j = 2$ this gives $\pi_1(1/2) + \pi_3(3/5) = \pi_2$, so $\pi_1(1/2) = \pi_2 - \pi_3(3/5)$, so $\pi_1 = 2\pi_2 - (6/5)\pi_3$. Combining the two equations, $\pi_2(1/3) + \pi_3(2/5) = 2\pi_2 - (6/5)\pi_3$, so $\pi_3(8/5) = \pi_2(2 - (1/3)) = (5/3)\pi_2$, and $\pi_3 = (5/8)(5/3)\pi_2 = (25/24)\pi_2$. Then $\pi_1 = 2\pi_2 - (6/5)\pi_3 = 2\pi_2 - (6/5)(25/24)\pi_2 = 2\pi_2 - (5/4)\pi_2 = (3/4)\pi_2$. We need $\pi_1 + \pi_2 + \pi_3 = 1$, i.e. $(3/4)\pi_2 + \pi_2 + (25/24)\pi_2 = 1$, i.e. $(67/24)\pi_2 = 1$. So, $\pi_2 = 24/67$. Then $\pi_1 = (3/4)\pi_2 = (3/4)(24/67) = 18/67$, and $\pi_3 = (25/24)\pi_2 = (25/24)(24/67) = 25/67$. So, the stationary distribution is $\pi = (18/67, 24/67, 25/67)$.

2.2.3(a): We need $\pi_i p_{ij} = \pi_j p_{ji}$ for all $i, j \in S$. So, $\pi_1 p_{12} = \pi_2 p_{21}$, i.e. $pi_1(1/3) = \pi_2(2/3)$, so $\pi_1 = 2\pi_2$. And, $\pi_2 p_{23} = \pi_3 p_{32}$, i.e. $pi_2(1/3) = \pi_3(2/3)$, so $\pi_2 = 2\pi_3$. And, $\pi_3 p_{34} = \pi_4 p_{43}$, i.e. $pi_3(1/3) = \pi_4(2/3)$, so $\pi_3 = 2\pi_4$. Also $\pi_1 + \pi_2 + \pi_3 + \pi_4 = 1$, so we must have $\pi = (8/15, 4/15, 2/15, 1/15)$. Then $\pi_i p_{ij} = \pi_j p_{ji}$ when $j = i + 1$ or $j = i - 1$ by the above, and when $j = i$ automatically, and when $|j - i| \geq 2$ since then both sides are 0. So, the chain is reversible with respect to π.

2.2.3(b): *This follows immediately from part (a) and Proposition (2.2.2).*

2.3.9: *Yes, it's aperiodic: since $p_{ii} > 0$ for all $i \in S$, therefore every state i has period 1.*

2.4.11(a): $p_{14}^{(2)} = \sum_{k \in S} p_{1k} p_{k4} = p_{11} p_{14} + p_{12} p_{24} + p_{13} p_{34} + p_{14} p_{44} = (0.1)(0.2) + (0.2)(0.1) + (0.5)(0.4) + (0.2)(0.3) = 0.02 + 0.02 + 0.20 + 0.06 = 0.30 = 0.3.$

2.4.11(b): *No. For example, if $i = 1$ and $j = 2$, then $\pi_i p_{ij} = (1/4)(0.2) = 1/20$, while $\pi_j p_{ji} = (1/4)(0.4) = 1/10$, so $\pi_i p_{ij} \neq \pi_j p_{ji}$.*

2.4.11(c): *Yes. The matrix P has each column-sum equal to 1 (as well as, of course, having each row sum equal to 1), i.e. it is doubly-stochastic. Hence, for any $j \in S$, we have that $\sum_{i \in S} \pi_i p_{ij} = \sum_{i \in S} (1/4) p_{ij} = (1/4) \sum_{i \in S} p_{ij} = (1/4)(1) = 1/4 = \pi_j$. So, π is a stationary distribution.*

2.4.11(d): *Yes. P is irreducible (since $p_{ij} > 0$ for all $i, j \in S$), and aperiodic (since $p_{ii} > 0$ for some, in fact all, $i \in S$), and π is stationary by part (c) above, so by the Markov Chain Convergence Theorem, we have $\lim_{n \to \infty} p_{ij}^{(n)} = \pi_j$ for all $i, j \in S$.*

2.4.12(a): *Yes, it's aperiodic: since $p_{ii} > 0$ for all $i \in S$ and it is irreducible, therefore every state i has period 1.*

2.4.12(b): *We need $\pi_1 + \pi_2 = 1$, and also $\pi_1 p_{11} + \pi_2 p_{21} = \pi_1$ and $\pi_1 p_{12} + \pi_2 p_{22} = \pi_2$. The second equation gives $\pi_1(2/3) + \pi_2(1/4) = \pi_1$, i.e. $\pi_2(1/4) = \pi_1(1/3)$, i.e. $\pi_2 = (4/3)\pi_1$. Then since $\pi_1 + \pi_2 = 1$, $(4/3)\pi_1 + \pi_1 = 1$, so $(7/3)\pi_1 = 1$, so $\pi_1 = 3/7$, and then $\pi_2 = 4/7$. Then all the equations are satisfied.*

2.4.12(c): *Yes it does. The chain is irreducible, aperiodic, and has a stationary distribution $\{\pi_i\}$, so by the Markov Chain Convergence Theorem, $\lim_{n \to \infty} p_{ij}^{(n)} = \pi_j$ for all $i, j \in S$. Setting $i = 1$ and $j = 2$ gives the result.*

2.4.13(a): *Here $S = \{1, 2, 3, 4, 5, 6, 7, 8, 9, 10\}$, and $\nu_1 = 1$ (with $\nu_i = 0$ for all other i). Also, for $1 \le i \le 9$, $p_{i,i+1} = 1/4$, and for $1 \le i \le 8$, $p_{i,i+2} = 3/4$, and $p_{10,1} = 1/4$, and $p_{9,1} = p_{10,2} = 3/4$, with $p_{ij} = 0$ otherwise.*

2.4.13(b): *Yes, since it is always possible to move one space clockwise, and hence eventually get to every other state with positive probability.*

2.4.13(c): *From any state i, it is possible to return in 10 seconds by moving one pad clockwise at each jump, or to return in 9 seconds by moving two pads clockwise on the first jump and then one pad clockwise for 8 additional jumps. Since $gcd(10, 9) = 1$, the chain is aperiodic.*

2.4.13(d): *Since the chain is irreducible, and the state space is finite, by the Finite Space Theorem, $\sum_{n=1}^{\infty} p_{ij}^{(n)} = \infty$ for all $i, j \in S$, so $\sum_{n=1}^{\infty} p_{15}^{(n)} = \infty$.*

2.4.13(e): *For every state $j \in S$, $\sum_{i \in S} p_{ij} = (1/4) + (3/4) = 1$. Hence, the chain is <u>doubly stochastic</u>. So, since $|S| < \infty$, the <u>uniform</u> distribution on S is a stationary distribution. Hence, we can take $\pi_1 = \pi_2 = \ldots = \pi_{10} = 1/10$.*

2.4.13(f): *Yes, since the chain is irreducible and aperiodic with stationary distribution $\{\pi_i\}$, therefore by the Markov Chain Convergence Theorem, $\lim_{n \to \infty} p_{ij}^{(n)} = \pi_j = 1/10$ for all $i, j \in S$, so $\lim_{n \to \infty} p_{15}^{(n)} = \pi_5 = 1/10$ exists.*

2.4.14(a): *Exists. For example, let $S = \{1, 2\}$, with $p_{12} = p_{21} = 1$ (and $p_{11} = p_{22} = 0$). Then the chain is irreducible (since it can get from each i to $3 - i$ in one step, and from i to i in two steps), and periodic with period 2 (since it only returns to each i in even numbers of steps). Furthermore, if $\pi_1 = \pi_2 = 1/2$, then for $i \neq j$, we have $\pi_i p_{ij} = (1/2)(1) = \pi_j p_{ji}$. Hence, the chain is reversible with respect to π, so π is a stationary distribution.*

2.4.14(b): *Does not exist. By the Equal Periods Lemma, since the chain is irreducible, all states must have the same period.*

2.4.14(c): *Does not exist. If the chain is reversible with respect to π, then π is a stationary distribution. Then if it is also irreducible, then by the Stationary Recurrence Theorem, it is recurrent, i.e. it is not transient.*

2.4.14(d): *Exists. For example, let $S = \{1, 2, 3\}$, with $p_{12} = p_{23} = p_{31} = 1/3$, and $p_{21} = p_{32} = p_{13} = 2/3$ (with $p_{ij} = 0$ otherwise).*

And let $\pi_1 = \pi_2 = \pi_3 = 1/3$, so π is a probability distribution on S. Then $\pi_1 p_{12} = (1/3)(1/3) \neq (1/3)(2/3) = \pi_2 p_{21}$, so the chain is not reversible with respect to π. On the other hand, for any $j \in S$, we have $\sum_i \pi_i p_{ij} = (1/3)(1/3 + 2/3) = 1/3 = \pi_j$, so π is a stationary distribution.

2.4.14(e): Exists. For example, let $S = \{1, 2, 3, 4, 5, 6\}$, with $p_{12} = p_{23} = p_{34} = p_{56} = p_{61} = 1$ (with $p_{ij} = 0$ otherwise). Let $i = 1$, and $j = 2$, and $k = 4$. Then $p_{ij} = p_{12} = 1 > 0$, and $p_{jk}^{(2)} = p_{23} p_{34} = 1 > 0$, and $p_{ki}^{(3)} = p_{45} p_{56} p_{61} = 1 > 0$, but state i has period 6 since it is only possible to return from i to i in multiples of six steps.

2.4.16(a): Possible. For example, let $S = \{1, 2, 3\}$, with $p_{12} = p_{23} = p_{31} = 1$ (and $p_{ij} = 0$ otherwise). Then the chain is irreducible (since it can get from $1 \to 2 \to 3 \to 1$), and periodic with period 3 (since it only returns to each i in multiples of three steps). Furthermore the chain is doubly stochastic, so if $\pi_1 = \pi_2 = \pi_3 = 1/3$, then π is a stationary distribution.

2.4.16(b): Possible. For example, let $S = \{1, 2, 3, 4, 5, 6\}$, with $p_{12} = p_{21} = 1$, and with $p_{34} = p_{45} = p_{56} = p_{63} = 1$. Then state $k = 1$ has period 2 since it only returns in multiples of 2 steps, and state $\ell = 3$ has period 4 since it only returns in multiples of 4 steps. (Of course, this chain is not irreducible; for irreducible chains, all states must have the same period.)

2.4.16(c): Possible. For example, let $S = \{1, 2, 3\}$, with $p_{12} = p_{23} = p_{31} = 1/3$, and $p_{21} = p_{32} = p_{13} = 1/2$, and $p_{11} = p_{22} = p_{33} = 1/6$. Then $0 < p_{ij} < 1$ for all $i, j \in S$ (yes, even when $i = j$). Next, let $\pi_1 = \pi_2 = \pi_3 = 1/3$, so π is a probability distribution on S. Then $\pi_1 p_{12} = (1/3)(1/3) \neq (1/3)(1/2) = \pi_2 p_{21}$, so the chain is not reversible with respect to π. On the other hand, for any $j \in S$, we have $\sum_i \pi_i p_{ij} = (1/3)(1/3 + 1/2 + 1/6) = 1/3 = \pi_j$. (Or, alternatively, $\sum_i p_{ij} = 1/3 + 1/2 + 1/6 = 1$, so the chain is doubly stochastic.) Hence, π is a stationary distribution.

2.4.16(d): Not possible. If the chain is irreducible, and $\sum_{n=1}^{\infty} p_{k\ell}^{(n)} = \infty$, then by the Recurrence Equivalences Theorem, we must have $f_{ij} = 1$ for all i and j.

2.4.16(e): *Possible. For example, let* $S = \{1, 2, 3, 4, 5, 6\}$, *with* $p_{12} = p_{15} = 1/2$, *and* $p_{23} = p_{34} = p_{45} = p_{56} = p_{61} = 1$, *with* $p_{ij} = 0$ *o.w. Let* $i = 1$, *and* $j = 2$, *and* $k = 4$. *Then* $p_{ij} = p_{12} = 1/2 > 0$, *and* $p_{jk}^{(2)} = p_{23}p_{34} = 1(1) = 1 > 0$, *and* $p_{ki}^{(3)} = p_{45}p_{56}p_{61} = 1(1)(1) = 1 > 0$, *but state* i *has period 3 (which is odd) since from* i *the chain can return to* i *in three steps* $(1 \to 5 \to 6 \to 1)$ *or six steps* $(1 \to 2 \to 3 \to 4 \to 5 \to 6 \to 1)$, *and* $\gcd(3, 6) = 3$.

2.4.16(f): *Not possible. If* $f_{ij} > 0$, *then there is some* $n \in \mathbf{N}$ *with* $p_{ij}^{(n)} > 0$.

2.4.16(g): *Possible. For example, let* $S = \{1, 2, 3, 4\}$, *and* $i = 1$ *and* $j = 2$, *and* $p_{12} = p_{13} = 1/2$, *and* $p_{24} = p_{34} = p_{44} = 1$. *Then* $p_{ij}^{(1)} = 1/2$, *but* $p_{ij}^{(n)} = 0$ *for all* $n \geq 2$, *so* $f_{ij} = 1/2$.

2.4.16(h): *Possible. For example, simple random walk with* $p = 3/4$ *has all states transient, but also* $f_{ij} > 0$ *and* $f_{ji} > 0$ *for all states* i *and* j *by irreducibility.*

2.4.16(i): *Not possible. One way to eventually get from* i *to* k, *is to first eventually get from* i *to* j, *and then eventually get from* j *to* k. *This means we must have* $f_{ik} \geq f_{ij} f_{jk} = (1/2)(1/3) = 1/6$, *so we cannot have* $f_{ik} = 1/10$.

2.4.19: *Yes. Here* π *is stationary by Problem (2.1.2)(c), and the chain is irreducible since e.g. it can go* $1 \to 2 \to 3 \to 2 \to 1$, *and the chain is aperiodic since e.g. it can get from 1 to 1 in two steps* $(1 \to 2 \to 1)$ *or three steps* $(1 \to 2 \to 3 \to 1)$ *and* $\gcd(2, 3) = 1$. *Hence, by the Markov chain Convergence Theorem,* $\lim_{n \to \infty} p_{ij}^{(n)} = \pi_j$ *for all* $i, j \in S$.

2.4.20(a): *Here* $S = \{1, 2, 3, 4, 5, 6, 7, 8, 9, 10\}$, *and* $\nu_1 = 1$ *(with* $\nu_i = 0$ *for all other* i). *Also, for* $1 \leq i \leq 9$, $p_{i,i+1} = 1/4$, *and for* $1 \leq i \leq 8$, $p_{i,i+2} = 3/4$, *and* $p_{10,1} = 1/4$, *and* $p_{9,1} = p_{10,2} = 3/4$, *with* $p_{ij} = 0$ *otherwise.*

2.4.20(b): *Yes, since it is always possible to move one space clockwise, and hence eventually get to every other state with positive probability.*

2.4.20(c): *From any state* i, *it is possible to return in 10 seconds by moving one pad clockwise at each jump, or to return in 9 seconds* *by moving two pads clockwise on the first jump and then one pad clockwise for 8 additional jumps. Since $gcd(10, 9) = 1$, the chain is aperiodic.*

2.4.20(d): *Since the chain is irreducible, and the state space is finite, by the Finite Space Theorem we have $\sum_{n=1}^{\infty} p_{ij}^{(n)} = \infty$ for all $i, j \in S$, so in particular $\sum_{n=1}^{\infty} p_{23}^{(n)} = \infty$.*

2.4.20(e): *For every state $j \in S$, $\sum_{i \in S} p_{ij} = (1/4) + (3/4) = 1$. Hence, the chain is <u>doubly stochastic</u>. So, since $|S| < \infty$, the <u>uniform</u> distribution on S is a stationary distribution. Hence, we can take $\pi_1 = \pi_2 = \ldots = \pi_{15} = 1/10$.*

2.4.20(f): *Yes, since the chain is irreducible and aperiodic with stationary distribution $\{\pi_i\}$, therefore $\lim_{n \to \infty} p_{ij}^{(n)} = \pi_j = 1/15$ for all $i, j \in S$, and in particular $\lim_{n \to \infty} p_{23}^{(n)} = \pi_3 = 1/10$.*

2.4.20(g): *Since $\lim_{n \to \infty} p_{15}^{(n)} = 1/10$, therefore also $\lim_{n \to \infty} p_{15}^{(n+1)} = 1/10$, and hence also $\lim_{n \to \infty} \frac{1}{2}[p_{15}^{(n)} + p_{15}^{(n+1)}] = 1/10$.*

2.5.5(a): *Here $\pi_1 p_{12} = (1/8)(1) = (1/2)(1/4) = \pi_2 p_{21}$, and $\pi_1 p_{13} = (1/8)(0) = (3/8)(0) = \pi_3 p_{31}$, and $\pi_3 p_{32} = (3/8)(1) = (1/2)(3/4) = \pi_2 p_{23}$, so $\pi_i p_{ij} = \pi_j p_{ji}$ for all $i, j \in S$, so the chain is reversible with respect to π.*

2.5.5(b): *Here π is stationary by part (a), and the chain is irreducible since it can go $1 \to 2 \to 3 \to 2 \to 1$, but the chain has period 2 since it always moves from odd to even or from even to odd. Hence, $p_{11}^{(n)} = 0$ whenever n is odd, so we do <u>not</u> have $\lim_{n \to \infty} p_{11}^{(n)} = 1/8$. But by the Periodic Convergence Theorem (2.5.1), we <u>do</u> still have $\lim_{n \to \infty} \frac{1}{2}[p_{11}^{(n)} + p_{11}^{(n+1)}] = \pi_1 = 1/8$, and by Average Probability Convergence (2.5.2) we also have $\lim_{n \to \infty} \frac{1}{n} \sum_{\ell=1}^{n} p_{11}^{(\ell)} = \pi_1 = 1/8$. So, in summary, (i) does <u>not</u> hold, but (ii) and (iii) <u>do</u> hold.*

2.6.2: *The Metropolis algorithm says that for $i \neq j$ we want $p_{ij} = (1/2) \min(1, \pi_j/\pi_i)$. So, we set $p_{21} = p_{32} = (1/2)(1) = 1/2$, and $p_{12} = (1/2)(2/3) = 1/3$ and $p_{23} = (1/2)(3/6) = 1/4$. Then to make $\sum_j p_{ij} = 1$ for all $i \in S$, we set $p_{11} = 2/3$, and $p_{22} = 1/4$, and $p_{33} = 1/2$. Then by construction, $\pi_i p_{ij} = \pi_j p_{ji}$ for all $i, j \in$*

S. Hence, the chain is reversible with respect to π. *Hence,* π *is a stationary distribution.*

2.7.12: *The graph is connected (since we can get from* $1 \rightarrow 2 \rightarrow 3 \rightarrow 4$ *and back), so the walk is irreducible. Also, the walk is aperiodic since e.g. we can get from 2 to 2 in 2 steps by* $2 \rightarrow 3 \rightarrow 2$, *or in 3 steps by* $2 \rightarrow 3 \rightarrow 4 \rightarrow 2$, *and* $\gcd(2,3) = 1$. *And, if* $\pi_u = d(u)/Z = d(u)/2|E| = d(u)/8$, *then* π *is a stationary distribution by (2.7.7). Hence,* $\lim_{n \to \infty} p_{21}^{(n)} = \pi_1 = d(1)/8 = 1/8$, *since* $d(1) = 1$ *because there is only one edge originating from the vertex 1.*

2.8.7(a):

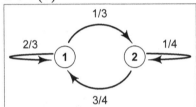

2.8.7(b): *From state 1, the chain returns immediately with probability 2/3, or with probability 1/3 moves to state 2 after which it returns in a Geometric(3/4) number of steps with mean* $1/(3/4) = 4/3$. *So,* $m_1 = 1 + (1/3)(4/3) = 13/9$. *Similarly, from state 2, the chain returns immediately with probability 1/4, or with probability 3/4 moves to state 2 after which it returns in a Geometric(1/3) number of steps with mean* $1/(1/3) = 3$. *So,* $m_2 = 1 + (3/4)(3) = 13/4$.

2.8.7(c): *By Theorem (2.8.5), we have* $\pi_1 = 1/m_1 = 9/13$, *and* $\pi_2 = 1/m_2 = 4/13$.

2.8.7(d): *Clearly* $p_{ii} \geq 0$, *and* $\sum_i \pi_i = (9/13)+(4/13) = 1$. *Also,* $\sum_i \pi_i p_{i1} = (9/13)(2/3)+(4/13)(3/4) = (18/39)+(12/52) = (6/13)+(3/13) = 9/13 = \pi_1$, *and* $\sum_i \pi_i p_{i2} = (9/13)(1/3) + (4/13)(1/4) = (9/39) + (4/52) = (3/13) + (1/13) = 4/13 = \pi_2$. *So,* π *is indeed a stationary distribution.*

2.9.1: *We again construct a Markov chain* $\{X_n\}$ *indicating how far along we are in the pattern at time* n. *This time* $S = \{0, 1, 2\}$, *with* $p_{01} = p_{12} = p_{21} = 1/2$ *for getting Heads, and* $p_{00} = p_{10} = p_{20} = 1/2$ *for getting Tails, so* $P = \begin{pmatrix} 1/2 & 1/2 & 0 \\ 1/2 & 0 & 1/2 \\ 1/2 & 1/2 & 0 \end{pmatrix}$. *We then* compute (check!) that this chain has stationary distribution $\pi = (1/2, 1/3, 1/6)$, so $\pi_2 = 1/6$, and then the waiting time for the sequence HH is equal to $1/\pi_2 = 1/(1/6) = 6$.

3.1.4: We need $\sum_{j \in S} j \, p_{ij} = i$ for each i. For $i = 1$ and $i = 4$ this is immediate, and for $i = 2$ it follows since $(1)(1/2) + (3)(1/2) = 2$, so it remains to consider $i = 3$. If $p_{31} = c$, then $p_{34} = 1 - c$, so $\sum_{j \in S} j \, p_{3j} = (1)(c) + (4)(1 - c) = 4 - 3c$. This equals 3 if $4 - 3c = 3$, i.e. $c = 1/3$, whence $p_{31} = 1/3$ and $p_{34} = 2/3$.

3.2.7(a): Intuitively, from 2, the chain has probability $1/3$ of decreasing by 1, so to preserve expectation it must have probability $1/6$ of increasing by 2, so $p_{24} = 1/6$. Similarly, from 3, the chain has probability $1/4$ of increasing by 1, so to preserve expectation it must have probability $1/4$ of decreasing by 1, so $p_{32} = 1/4$. More formally, for a Markov chain to be a martingale, we need that $\sum_{j \in S} j \, p_{ij} = i$ for all $i \in S$. With $i = 2$, we need that $(1/3)(1) + p_{22}(2) + p_{24}(4) = 2$, so $p_{22}(2) + p_{24}(4) = 5/3$, so $p_{22} + 2p_{24} = 5/6$. But also $p_{21} + p_{22} + p_{24} = 1$, whence $p_{22} + p_{24} = 2/3$. Subtracting this equation from the previous one gives $p_{24} = (5/6) - (2/3) = 1/6$, hence $p_{22} = (2/3) - (1/6) = 1/2$. Similarly, with $i = 3$, we need $p_{32}(2) + p_{33}(3) + (1/4)(4) = 3$, so $2p_{32} + 3p_{33} = 2$. But also $p_{32} + p_{33} + p_{34} = 1$, whence $p_{32} + p_{33} = 3/4$. Subtracting twice this equation from the previous one gives $p_{33} = 2 - 2(3/4) = 1/2$, hence $p_{32} = 1/4$. In summary, if $p_{24} = 1/6$, $p_{22} = 1/2$, $p_{32} = 1/4$, and $p_{33} = 1/2$, then we have valid Markov chain transitions which make it a martingale.

3.2.7(b): Clearly the chain is bounded up to time T, indeed we always have $|X_n| \leq 4$. Hence, by the Optional Stopping Corollary, $\mathbf{E}(X_T) = \mathbf{E}(X_0) = 3$.

3.2.7(c): Let $p = \mathbf{P}(X_T = 1)$. Then since we must have $X_T = 1$ or 4, therefore $\mathbf{P}(X_T = 4) = 1 - p$, and $\mathbf{E}(X_T) = p(1) + (1 - p)(4) = 4 - 3p$. Solving and using part (b), we must have that $3 = 4 - 3p$, so $3p = 4 - 3 = 1$, whence $p = 1/3$.

3.2.8(a): For a Markov chain to be a martingale, we need that $\sum_{j \in S} j \, p_{ij} = i$ for all $i \in S$. With $i = 2$, we need that $(1/4)(1) + p_{22}(2) + p_{24}(4) = 2$. But we must have $\sum_{j \in S} p_{ij} = 1$, i.e. $(1/4) + p_{22} + p_{24} = 1$, i.e. $p_{22} = 3/4 - p_{24}$, so we must have $(1/4)(1) + (3/4 - $

$p_{24})(2) + p_{24}(4) = 2$, or $p_{24}(4 - 2) = 2 - 1/4 - 3/2 = 1/4$, so $p_{24} = (1/4)/2 = 1/8$. (Or, more simply, from 2 the chain has probability $1/4$ of decreasing by 1, so to preserve expectations it must have probability $1/8$ of increasing by 2.) Then $p_{22} = 3/4 - p_{24} = 3/4 - 1/8 = 5/8$. Similarly, with $i = 3$, we need $p_{32}(2) + p_{33}(3) + (1/5)(4) = 3$. For simplicity, since we must have $p_{32} + p_{33} + (1/5) = 1$, we can subtract 3 from each term, to get that $p_{32}(-1) + p_{33}(0) + (1/5)(1) = 0$, so $p_{32} = 1/5$, and then $p_{33} = 1 - 1/5 - p_{32} = 3/5$. In summary, if $p_{24} = 1/8$, $p_{22} = 5/8$, $p_{32} = 1/5$, and $p_{33} = 3/5$, then we have valid Markov chain transitions which make it a martingale.

3.2.8(b): Clearly the chain is bounded up to time T, indeed we always have $|X_n| \leq 4$. Hence, by the Optional Stopping Corollary, $\mathbf{E}(X_T) = \mathbf{E}(X_0) = 2$.

3.2.8(c): Let $p = \mathbf{P}(X_T = 1)$. Then since we must have $X_T = 1$ or 4, therefore $\mathbf{P}(X_T = 4) = 1 - p$, and $\mathbf{E}(X_T) = p(1) + (1 - p)(4) = 4 - 3p$. Solving and using part (b), we must have that $2 = 4 - 3p$, so $3p = 4 - 2 = 2$, whence $p = 2/3$.

3.2.9(a): For $i \geq 2$, we need $\sum_j j p_{ij} = i$, so $(i - 1)c + (i + 2)(1 - c) = i$, so $i + 2 - 3c = i$, so $2 = 3c$, so $c = 2/3$.

3.2.9(b): We need $\sum_{j \in S} j p_{1j} = 1$. But $\sum_{j \in S} j p_{1j} = \sum_{j=1}^{\infty} j p_{1j} = p_{11} + \sum_{j=2}^{\infty} j p_{ij} \geq p_{11} + \sum_{j=2}^{\infty} 2 p_{ij} = p_{11} + 2(1 - p_{11}) = 2 - p_{11}$. For this to equal 1, we need $p_{11} = 1$.

3.2.9(c): Since $\{X_n\}$ is a martingale, $\mathbf{E}(X_n) = \mathbf{E}(X_0) = 5$ for all n, so in particular $\mathbf{E}(X_3) = 5$.

3.2.9(d): Clearly the chain is bounded up to time T, indeed we always have $|X_n| 1_{n \leq T} \leq 11$. Hence, by the Optional Stopping Corollary, $\mathbf{E}(X_T) = \mathbf{E}(X_0) = 5$.

3.2.9(e): Starting at 5, the chain has positive probability of immediately going $5 \to 4 \to 3 \to 2 \to 1$ and then getting stuck at 1 forever and never returning to 5. Hence, the state 5 is _transient_. It thus follows from the Recurrence Theorem that $\sum_{n=1}^{\infty} p_{55}^{(n)} < \infty$, i.e. $\sum_{n=1}^{\infty} p_{55}^{(n)} \neq \infty$. (Note that this chain is _not_ irreducible, since e.g. $f_{12} = 0$, so results like the Cases Theorem do _not_ apply.)

3.3.6: Let $r = (1 - p)/p$, so $p = 1/(r + 1)$, and $2p - 1 = 2/(r + 1) - 1 = (1 - r)/(1 + r)$. Then the formula for $\mathbf{E}(T)$ when

$p \neq 1/2$ can be written as

$$\frac{c\frac{r^a-1}{r^c-1} - a}{(1-r)/(1+r)}.$$

As $p \to 1/2$, we have $r \to 1$. So, we compute (using L'Hôpital's Rule twice, and the fact that as $r \to 1$, $1 + r \to 2$) that

$$\lim_{r \to 1} \frac{c\frac{r^a-1}{r^c-1} - a}{(1-r)/(1+r)} = 2 \lim_{r \to 1} \frac{c\frac{r^a-1}{r^c-1} - a}{1-r}$$

$$= 2 \lim_{r \to 1} \frac{c(r^a - 1) - a(r^c - 1)}{(r^c - 1)(1 - r)} = 2 \lim_{r \to 1} \frac{car^{a-1} - acr^{c-1}}{cr^{c-1}(1-r) - (r^c - 1)}$$

$$= 2ac \lim_{r \to 1} \frac{(a-1)r^{a-2} - (c-1)r^{c-2}}{c(c-1)r^{c-2}(1-r) - cr^{c-1} - cr^{c-1}}$$

$$= 2ac \frac{(a-1) - (c-1)}{c(c-1)1^{c-2}(1-1) - c1^{c-1} - c1^{c-1}}$$

$$= 2ac \frac{(a-c)}{0 - c - c} = 2ac \frac{(a-c)}{(-2c)} = a(c-a).$$

3.4.1: We suppose that at each time n, a player appears and bets \$1 on Tails, then if they win they bet \$2 on Tails, then if they win they bet \$ on Tails, otherwise they stop betting. Then players 1 through $\tau - 3$ receive either -1, or $+1 - 2$, or $+1 + 2 - 4$, so they always lose \$1. Player $\tau - 2$ wins all three bets and thus wins \$7. Player $\tau - 1$ wins two bets and thus wins \$3. Player τ wins one bet and thus wins \$1. So, $X_\tau = -(\tau - 3) + 7 + 3 + 1 = -\tau + 14$. But as before $\mathbf{E}(X_\tau) = 0$, so $\mathbf{E}(-\tau + 14) = 0$, i.e. $\mathbf{E}(\tau) = 14$.

3.5.4(a): Here $|X_n| \leq 5 + n$ so $\mathbf{E}|X_n| < \infty$. Also $\sum_j j\, p_{5j} = 5(1) + 0 = 5$, and for $i \geq 6$, $\sum_j j\, p_{ij} = (i - 1)(1/2) + (i + 1)(1/2) = i$, so $\{X_n\}$ is a martingale.

3.5.4(b): Since $\{X_n\}$ is a martingale which is clearly bounded below (by 5), the Martingale Convergence Theorem (3.5.3) says $X_n \to X$ for some random variable X. But if X_n reaches any other value i besides 5, then it will continue to oscillate and have probability 0 of remaining at (and hence converging to) i. So, the only remaining possibility is that $\mathbf{P}(X = 5) = 1$, i.e. that the chain converges to 5.

3.6.2(a): Here $X_1 = 0$ iff the first individual has 0 offspring, so $\mathbf{P}(X_1 = 0) = \mu\{0\} = 4/7$.

3.6.2(b): Here $X_1 = 0$ iff the first individual has 2 offspring, so $\mathbf{P}(X_1 = 2) = \mu\{2\} = 1/7$.

3.6.2(c): Here $X_2 = 2$ iff we have either: (i) the first individual has 1 offspring who then has 2 offspring, with probability $\mu\{1\}\mu\{2\} = (2/7)(1/7) = 2/49 = 14/343$, or (ii) the first individual has 2 offspring who then have 1 offspring each, with probability $\mu\{2\}\mu\{1\}\mu\{1\} = (1/7)(2/7)(2/7) = 4/343$, or (iii) the first individual has 2 offspring who then have 0 and 2 offspring respectively, in either order, with probability $\mu\{2\}(2)\mu\{0\}\mu\{2\} = (1/7)(2)(4/7)(1/7) = 8/343$, the first individual has 2 offspring, so $\mathbf{P}(X_2 = 2) = (14 + 4 + 8)/343 = 26/343$.

3.7.3: If the option price is c, and we buy x units of the stock, and y units of the option, then (a) if the stock is worth \$40 tomorrow, our profit is $x(40-50)+y(40-70)^+ -cy = -10x-cy$; while (b) if the stock is worth \$80 tomorrow, our profit is $x(80-50)+y(80-70)^+ - cy = 30x+(10-c)y$. If these are equal, then $-40x = 10y$, or $y = -4x$. Our profit will then be $-10x - cy = -10x - c(-4x) = (-10+4c)x$. There is no arbitrage if our profit is 0, i.e. $-10+4c = 0$, so $c = 10/4 = \$2.50$.

3.7.5(a): Here if you buy x stock and y options, then your profit if the stock goes down is $x(-10) + y(-c)$, or if the stock goes up is $x(30) + y(20 - c)$. These are equal if $y = -2x$, in which case they both equal $-10x+2xc = 2x(c-5)$. So, there is no arbitrage iff $c = 5$, i.e. the fair price is \$5.

3.7.5(b): The stock price is a martingale if the stock price goes down with probability $3/4$, or up with probability $1/4$. And, under those martingale probabilities, the expected value of the option tomorrow is $(3/4)(0) + (1/4)(20) = 5 = c$. So, again, the fair price for the option is \$5. (And once again, the _true_ probabilities that the stock tomorrow equals \$10 or \$50 are irrelevant.)

3.7.8: In this case, $a = 20$, $d = 10$, $u = 50$, and $K = 30$. So, the fair price $= c = (20 - 10)(50 - 30)/(50 - 10) = 5$. Alternatively, the martingale probabilities are $q_1 = (50 - 20)/(50 - 10) = 3/4$ and $q_2 = 1 - q_1 = 1/4$, so the fair price = martingale expected value = $q_1(0) + q_2(u - K) = (1/4)(50 - 30) = 5$. Either way, it gives the same answer as before.

4.1.2: Here $\mathbf{E}[(B_2 + B_3 + 1)^2] = \mathbf{E}[(B_2)^2] + \mathbf{E}[(B_3)^2] + 1^2 + 2\mathbf{E}[B_2 B_3] + 2\mathbf{E}[B_2(1)] + 2\mathbf{E}[B_3(1)] = 2 + 3 + 1 + 2(2) + 2(0) + 2(0) = 10$.

4.1.3: $\mathrm{Var}[B_3 + B_5 + 6] = \mathbf{E}([(B_3 + B_5 + 6) - \mathbf{E}(B_3 + B_5 + 6)]^2) = \mathbf{E}([(B_3 + B_5 + 6) - (6)]^2) = \mathbf{E}([B_3 + B_5]^2) = \mathbf{E}([B_3^2]) + \mathbf{E}([B_5]^2) + 2\mathbf{E}(B_3 B_5) = 3 + 5 + 2(3) = 14$.

4.2.6: We compute by plugging into the formula (4.2.4) that the fair prices are (i) \$10.12, and (ii) \$13.21. The second price is larger, because the volatility is larger while all other parameters remain the same, and options help to protect against volatility.

4.3.6(a): $N(3) \sim \mathrm{Poisson}(3\lambda) = \mathrm{Poisson}(12)$, so $\mathbf{P}[N(3) = 5] = e^{-12}(12)^5/5!$.

4.3.6(b): $\mathbf{P}[N(3) = 5 \mid N(2) = 1] = \mathbf{P}[N(3) = 5, \ N(2) = 1] / \mathbf{P}[N(2) = 1] = \mathbf{P}[N(2) = 1, \ N(3) - N(2) = 4] / \mathbf{P}[N(2) = 1] = \mathbf{P}[N(2) = 1]\,\mathbf{P}[N(3) - N(2) = 4] / \mathbf{P}[N(2) = 1] = \mathbf{P}[N(3) - N(2) = 4] = e^{-4}4^4/4!$.

4.3.6(c): $\mathbf{P}[N(3) = 5 \mid N(2) = 1, N(6) = 6] = \mathbf{P}[N(3) = 5, \ N(2) = 1, \ N(6) = 6] / \mathbf{P}[N(2) = 1, \ N(6) = 6] = \mathbf{P}[N(2) = 1, \ N(3) - N(2) = 4, \ N(6) - N(3) = 1] / \mathbf{P}[N(2) = 1, \ N(6) - N(2) = 5] = \mathbf{P}[N(2) = 1]\,\mathbf{P}[N(3) - N(2) = 4]\,\mathbf{P}[N(6) - N(3) = 1] / \mathbf{P}[N(2) = 1]\,\mathbf{P}[N(6) - N(2) = 5] = \mathbf{P}[N(3) - N(2) = 4]\,\mathbf{P}[N(6) - N(3) = 1] / \mathbf{P}[N(6) - N(2) = 5] = [e^{-4}4^4/4!]\,[e^{-12}12^1/1!] / [e^{-16}16^5/5!] = [4^4/4!]\,[12]/[16^5/5!]$.

4.4.12: For (a), we need $\pi G = 0$, i.e. $-3\pi_1 + \pi_2 = 0$, and $3\pi_1 - 2\pi_2 + 4\pi_3 = 0$, and $\pi_2 - 4\pi_3 = 0$. From the first equation, $\pi_1 = (1/3)\pi_2$, and from the third, $\pi_3 = (1/4)\pi_2$. Since $\pi_1 + \pi_2 + \pi_3 = 1$, we must have $\pi = (4/19,\ 12/19,\ 3/19)$. For (b), we need $\pi_i g_{ij} = \pi_j g_{ji}$ for all i and j. Taking $i = 1$ and $j = 2$ gives $\pi_1(3) = \pi_2(1)$ so $\pi_1 = (1/3)\pi_2$. Taking $i = 2$ and $j = 3$ gives $\pi_2(1) = \pi_3(4)$ so $\pi_3 = (1/4)\pi_2$. So, this again implies that $\pi = (4/19,\ 12/19,\ 3/19)$.

4.5.2(a): To get from 5 to 7 requires <u>either</u> at least three arrivals and one departure with probability $o(h^2)$, <u>or</u> two arrivals and zero departures with probability $\left(e^{-\lambda h}(\lambda h)^2/2!\right)\left(e^{-\mu h}(\lambda h)^0/0!\right) = \lambda^2 h^2/2 + o(h^2)$. This is $O(h^r)$ if $r = 2$.

4.5.2(b): With $r = 2$, the above says that $\lim_{h\searrow 0}[p_{57}(h)\,/\,h^r] = \lim_{h\searrow 0}[(\lambda^2 h^2/2 + o(h^2))\,/\,h^2] = \lambda^2/2 = 6^2/2 = 18$.

4.5.8: Here $\mathbf{E}(Y_n) = 1/5$ and $\mathbf{E}(S_n) = C/2$. So, $Q(t)$ is bounded in probability iff $\mathbf{E}(S_n) < \mathbf{E}(Y_n)$, i.e. $C/2 < 1/5$, i.e. $C < 2/5$.

4.6.8(a): Here μ is the mean of the Uniform$[10, 20]$ distribution, so $\mu = (10+20)/2 = 15$. Then, by the Elementary Renewal Theorem, $\lim_{t\to\infty}[N(t)/t] = 1/\mu = 1/15$.

4.6.8(b): Here $h = 6$ is small enough that $\mathbf{P}(Y_n < 6) = 0$. So, by Corollary (4.6.7), $\lim_{t\to\infty}\mathbf{P}(\exists n \geq 1 : t < T_n < t + 6\}) = \lim_{t\to\infty}\mathbf{E}[N(t + h) - N(t)] = h/\mu = 6/15$.

4.7.10: Yes. This chain satisfies that $P(x, \cdot)$ has positive density function on $[x - \delta, x + \delta]$ where $\delta = 5 > 0$. So, the chain is ϕ-irreducible by Proposition (4.7.3), and is aperiodic by Proposition (4.7.6).

4.7.11(a): Yes. If we let $\phi = $ Uniform$[0, 1]$, then for any $x \in S$, whenever $\phi(A) > 0$, then also $P(x, A) > 0$. So, the chain is ϕ-irreducible for this choice of ϕ. (But <u>not</u> for the choice where ϕ is Lebesgue measure on $\mathbf{R}$.)

4.7.11(b): Yes. If to the contrary the period was $b \geq 2$, then there must be at least one S_i which has positive measure according to the Uniform$[0,1]$ distribution. For this S_i, we have $P(x, S_i) > 0$ for all $x \in S$, even for $x \in S_i$, which contradicts the forced cycles in Definition (4.7.4).

4.7.11(c): Yes. If $\pi = $ Uniform$[0, 1]$, then $P(x, \cdot) = \pi$ for all $x \in S$. So, if $X_0 \sim \pi$ (or even if it isn't!), then we have $\mathbf{P}(X_1 \in A) = \int P(x, A)\,\pi(dx) = \int \pi(A)\,\pi(dx) = \pi(A)$, so π is a stationary distribution.

4.7.11(d): Yes. Since the chain is ϕ-irreducible, and aperiodic, and has a stationary distribution π, it follows from the General Space Convergence Theorem (4.7.9) that $\lim_{n\to\infty}\mathbf{P}_x(X_n \in A) = \pi(A)$ for all $A \subseteq S$ and π-a.e. $x \in S$.

4.8.2: *The Metropolis algorithm first chooses X_0 from some initial distribution ν, say $X_0 = 5$. Then, for $n = 1, 2, \ldots$, given X_{n-1}, it first samples $U_n \sim \text{Uniform}[0, 1]$ and $Y_n \sim q(X_{n-1}, \cdot) = \text{Uniform}[X_{n-1} - 1/2, X_{n-1} + 1/2]$. Then, if $U_n \leq \pi(Y_n) / \pi(X_{n-1})$ then it sets $X_n = Y_n$ ("accept"), otherwise if $U_n > \pi(Y_n) / \pi(X_{n-1})$ then it sets $X_n = X_{n-1}$ ("reject"). It continues in this way for a large number M of steps, after which X_M approximately has the distribution π.*

Index (Note: Page references to figures are underlined.)

Printed in the United States
By Bookmasters

A FIRST LOOK AT STOCHASTIC PROCESSES

This textbook introduces the theory of stochastic processes, that is, randomness which proceeds in time. Using concrete examples like repeated gambling and jumping frogs, it presents fundamental mathematical results through simple, clear, logical theorems and examples. It covers in detail such essential material as Markov chain recurrence criteria, the Markov chain convergence theorem, and optional stopping theorems for martingales. The final chapter provides a brief introduction to Brownian motion, Markov processes in continuous time and space, Poisson processes, and renewal theory.

Interspersed throughout are applications to such topics as gambler's ruin probabilities, random walks on graphs, sequence waiting times, branching processes, stock option pricing, and Markov chain Monte Carlo (MCMC) algorithms.

The focus is always on making the theory as well-motivated and accessible as possible, to allow students and readers to learn this fascinating subject as easily and painlessly as possible.

World Scientific
www.worldscientific.com
11488 sc

ISBN 978-981-120-897-3 (pbk)

9 789811 208973